THERMODYNAMICS OF SURFACE PHENOMENA

THERMODYNAMICS OF SURFACE PHENOMENA

R.Kh. Dadashev

Viva Books

New Delhi | Mumbai | Chennai | Kolkata | Bangalore | Hyderabad | Kochi

First Indian Edition 2009

VIVA BOOKS PRIVATE LIMITED

- 4737/23 Ansari Road, Daryaganj, New Delhi 110 002
  E-mail: vivadelhi@vivagroupindia.net, Tel. 42242200
- 76, Service Industries, Shirvane, Sector 1, Nerul, Navi Mumbai 400 706
  E-mail: vivamumbai@vivagroupindia.net, Tel. 27721273, 27721274
- Jamals Fazal Chambers, 26 Greams Road, Chennai 600 006
  E-mail: vivachennai@vivagroupindia.net, Tel. 28290304, 28294241
- B-103, Jindal Towers, 21/1A/3 Darga Road, Kolkata 700 017
  E-mail: vivakolkata@vivagroupindia.net, Tel. 22836381, 22816713
- 7, Sovereign Park Apartments, 56-58, K. R. Road, Basavanagudi, Bangalore 560 004
  E-mail: vivabangalore@vivagroupindia.net, Tel. 26607409, 26607410
- 101-102 Mughal Marc Apartments, 3-4-637 to 641, Narayanguda, Hyderabad 500 029
  E-mail: vivahyderabad@vivagroupindia.net, Tel. 27564481, 27564482
- Beevi Towers, First Floor, SRM Road, Kaloor, Kochi 682 018, Kerala
  E-mail: vivakochi@vivagroupindia.net, Tel: 0484-2403055, 2403056

www.vivagroupindia.com

Published by arrangement with
Cambridge International Science Publishing
7, Meadow Walk, Great Abington
Cambridge CB1 6AZ, UK

ISBN-10: 81-309-1115-9
ISBN-13: 978-81-309-1115-1

Published by Vinod Vasishtha for Viva Books Private Limited, 4737/23, Ansari Road, Daryaganj, New Delhi-110 002. Printed and bound by Anand Sons, Delhi-110 092.

CONTENTS

INTRODUCTION

The results of the examination of the surface properties of metals and alloys are used widely in solving a number of important scientific and technical problems (production of superpure materials and amorphous alloys, synthesis of alloys with the given properties, formation of energy conversion systems, etc. [1–3]). The surface phenomena are also important in the processes of brazing and coating [2, 4], modification of alloys and sintering [5–7], surface separation substances, etc. [8].

The further development of the theory of liquid metals [9–11] and, in particular, investigations of the structure of the melt–saturated vapour interface are not possible without obtaining reliable experimental data on the surface tension (σ) of metallic alloys with generalisation of these data. Therefore, special attention is given to the theoretical and experimental examination of the surface tension and molar volumes of metals and alloys.

A large amount of experimental data have been compiled on the properties of pure metals and binary systems [12]. Theoretical equations have been derived [7, 13, 14] for describing the concentration dependence of the surface tension of a large number of binary systems using the phase equilibrium data. The temperature and concentration dependences of the surface tension of metals and binary alloys can be used to determine: adsorption of components, excess entropy, composition, minimum possible and effective thickness of the surface layer [6, 19].

However, insufficient attention has been given to the theoretical and experimental investigations of the surface properties of multicomponent systems. No exact theory has been proposed for the quantitative description of the dependence of surface tension on composition and temperature. The currently available theoretical equations either contain physical quantities which are very difficult to determine by experiments or often do not describe, even quantitatively, the surface properties of melts. Exact methods for calculating the thermodynamic parameters of the surface layer of multicomponent alloys on the basis of the concentration dependence of surface tension and molar volumes have not been proposed.

This situation is caused by the considerable difficulties in theoretical analysis of the processes taking place at the multicomponent solution-saturated vapour interface. On the other hand, the experimental examination of the properties of the surface of metallic melts is associated with considerable difficulties which become even more serious in transition from binary to multicomponent systems.

At the same time, the majority of materials used in practice are multicomponent alloys and, consequently, it is necessary to examine their properties in detail. When solving a large number of technological problems, it is important to take into account the effect of several components on the properties of materials, and the surface-active substances, even in very small quantities, can greatly influence the composition and properties of surface layers [6, 7]. Therefore, the task of investigating the surface properties of multicomponent systems is still very important.

In the middle of the 70s of the previous century, the Laboratory of the Physics of Interfacial Phenomena of the Chechen–Ingushsk State University started experimental and theoretical investigations of the surface properties of ternary and four-component systems under the supervision of Kh.I. Ibragimov. At that time, experimental data for the surface tension and density were available only for several ternary metallic systems. No investigations of the surface properties of systems with a large number of components had been carried out, with the exception of individual technical alloys. The aim of our research was to carry out systematic investigations of the surface properties (surface tension, adsorption, composition and effective thickness of the surface layer) of multicomponent metallic melts and to develop, using the results, simple and reliable methods for predicting using the properties of appropriate binary systems. The results were scattered through a large number of various sources, and most of them were difficult to obtain. Therefore, we believe that it was interesting to collect and present these results in a single book. It is up to the reader to say whether we have succeeded. The book was written over a number years with interruptions during the serious situation in the Chechen Republic..

The book has six chapters.

Chapter 1 examines the Gibbs method in the thermodynamics of surface phenomena. The concepts of the dividing surface and excess values in the Gibbs capillarity theory are investigated in the form that is easy to understand. Special attention is given to the dependence of excess quantities on the position of the dividing surface.

The adsorption phenomena in two-phase multicomponent systems are analysed. Equations have been derived and a method developed for calculating the excess entropy and adsorption of components in relation to different positions of the dividing surface on the temperature and concentration dependence of interfacial tension. Various variants of the change of the composition of multicomponent solutions are discussed.

The results of comparative analysis are presented and the methods of changing the composition most suitable for practice described. The general thermodynamic relationships are determined between the adsorption values in relation to different positions of the Gibbs dividing surface in the multicomponent two-phase system. The adsorption method of determination of the molecular mass of surface-active substances is described as an example.

The second chapter is concerned with the description of the surface by the method of the finite thickness layer. The method is used to derive the systems of equations for calculating the main thermodynamic parameters of the surface layer of the multicomponent melt (molar and partial–molar areas, composition and effective thickness of the surface layer). The thermodynamic equations, describing the concentration dependence of the surface tension of multicomponent solutions, are compared.

Experimental equipment for the measurement of the surface tension and density of liquid metals and alloys was designed and constructed for the experimental investigations of multicomponent systems. The existing devices were improved and a series of devices was produced for the combined determination of surface tension and density greatly increasing the productivity of experiments without reducing the accuracy of measurement. These problems, together with a detailed analysis of the measurement error of surface tension and density, are investigated in the third chapter.

The fourth and fifth chapters are concerned with the experimental data on the surface properties of binary and multicomponent systems.

The method of mathematical experiment design is used for the detailed investigations of the surface tension and density of binary (indium–tin, thallium–lead, thallium–bismuth, tin–thallium, indium-thallium, gallium–bismuth), ternary (indium–tin–lead, indium–tin-gallium, indium–tin–thallium, thallium–lead–bismuth, thallium–lead-indium, thallium–lead–tin, indium-lead-bismuth) and quaternary (indium-tin–lead–bismuth and indium–gallium–lead–bismuth) metallic melts. Using the results, some of the contradictions in the experimental results have been removed and new data have been obtained which make it possible to determine the main relationships of the variation of the surface tension, density and molar volumes with temperature and composition in binary, ternary and quaternary systems.

Using the experimental data, the methods which we proposed have been used to calculate the main thermodynamic parameters of the surface layer of the investigated melt (adsorption, composition and thickness of the surface layer, molar and partial–molar area of the components). Special features of the concentration and temperature dependences of these properties in the investigated systems have been determined.

It has been established that the special features of the concentration dependence of the surface property multicomponent metallic melts are linked directly with the isotherms of these properties in lateral binary systems. This is the subject of the sixth chapter.

The results are generalised in the sixth chapter. The results of the comparative analysis of the experimental data for the surface tension of multicomponent metallic melts have been used to develop and verify on a large number of experimentally investigated ternary and quaternary systems a simple method of predicting the surface tension of multicomponent melts on the basis of the properties of the binary systems which form these melts. The comparison of the calculated values with the experimental data showed the good agreement of these results in both the systems with smooth surface tension isotherms and with complicated isothermal surfaces.

1

The Gibbs method in the thermodynamics of surface phenomena

Two main methods of examining the interface are used in the modern thermodynamics of surface phenomena. The first method, proposed by Young and developed by Gibbs more than 100 years ago [15], is based on the concepts such as the dividing surface, the idealised reference system, and excess thermodynamic quantities. This method does not require any assumption to be made regarding the thickness of the interphase layer.

A real two-phase system in the method is replaced by a set of two phases with a strictly defined geometrical surface between them. The properties of the phases are assumed to be identical to the dividing surface (the Gibbs surface) on which all extensive parameters change abruptly.

The second method [16–20] examines the surface layer as an independent phase, with a finite thickness and restricted by two mathematical surfaces. Both methods have advantages and shortcomings which have been discussed in detail in monographs [6, 13, 20–23]. The application of a specific method depends on the ease of solving the given task [23].

Fundamental equations, derived by the method of the finite thickness layer, are analysed in detail in the second part. This chapter is concerned with the special features of application of the Gibbs theory for analysis of surface phenomena in multicomponent systems.

Regardless of the large number of investigations concerned with the analysis of the surface phenomena on the basis of the Gibbs fundamental thermodynamic equation, the task of calculating the adsorption of components from multicomponent solutions on the basis of the concentration dependence of surface tension has not been solved. Usually, examination of the problem is restricted to the application of the Gibbs adsorption equation for a binary solution. Equations for calculating the adsorption and composition of the surface layer of ternary systems have been derived in [6, 13, 20] and only one study [8] publishes a system of the equations for calculating relative Gibbs adsorption in multicomponent systems.

It is also believed [6, 25] that it is not possible to determine the adsorption in real multicomponent systems using the Gibbs adsorption equation, although Gibbs already showed that relative adsorption can also be calculated in this case [15]. Therefore, the development of methods of calculating adsorption on various dividing surfaces using the data on the concentration dependence of surface tension in real multicomponent systems is still an urgent task.

In Chapter 1, attention is given to the main assumptions of Gibbs capillarity theory with reference to multicomponent two-phase systems. Within the framework of the theory, equations are derived and a method is developed for calculating excess thermodynamic quantities using the dependence of surface tension or the temperature and composition of one of the co-existing phases. The adsorption of components at the boundary of two multicomponent fluid phases is also calculated. General thermodynamic relationships are established between the values of the adsorption of components on different dividing surfaces.

1.1. Dividing surface and excess quantity in Gibbs capillarity theory

Because of the brevity of the explanation typical of Gibbs, some of the assumptions of the theory of capillarity, as reported in the literature [19, 20, 26], are incorrectly perceived and this results in objections and misunderstanding [23, 27, 28]. Therefore, attention will initially be given to excess quantities, their dependence on the position of the dividing surface, and also to the concept of the dividing surface in Gibbs theory. This greatly facilitates further examination of more complicated problems.

The direct object of examination of the physics of surfaces is the anisotropic and heterogeneous, in one direction, zone of the continuous variation of the local properties, situated between coexisting phases. Gibbs referred to this zone as the fracture zone. Further, when examining the zone as a separate phase with a specific thickness, the terms the surface layer, the interfacial or transition zone were introduced. Sometimes, the zone is referred to as the transition layer or transition region.

Because of the small radius of action of intermolecular forces, the properties of the surface layer may change considerably already at intermolecular distances resulting in steep gradients of the properties and the concept of a jump in the properties (in the physical sense of the word) at the phase boundary. Therefore, the concept of the fracture surface has changed greatly with the development of the theory of surface phenomena. Evidently, the length of the zone should depend in particular on the nature of the bordering phases and the interface, on the intermolecular forces and the radius of action of these forces, and also on temperature and pressure.

We examine a multiphase, multicomponent system with fracture surfaces in the state of thermodynamic equilibrium. Examination will be restricted to flat surfaces.

The fracture surface is characterised by a gradient of the properties along the normal to the surface. It is therefore incorrect to apply the laws of equilibrium thermodynamics to this heterogeneous layer. These laws could be applied to the interfacial layer after Gibbs introduced the concept of the dividing surface and the idealised thermodynamic reference system. These concepts must be used in the examination of distorted interfaces. In the case of a flat boundary, as mentioned previously, the second method proposed by Guggenheim has been used with success.

In a letter to the American Academy written after receiving the Rumford prize, Gibbs wrote: 'one of the tasks of theoretical investigations in any area of knowledge is the establishment of a viewpoint from which the examined object can be viewed with simplicity [15, page 540]. As mentioned by Academician Rusanov [6], this viewpoint in Gibbs capillarity theory is the assumption on the dividing surface.

The dividing surface is the geometrical surface drawn hypothetically between the coexisting phases in the direction parallel to the fracture surface in such a manner that it reproduces the shape of the fracture surface. This concept is one of the most important concepts of the theory of capillarity. It is linked closely with the assumption on the idealised thermodynamic reference system.

The ideal thermodynamic reference system in Gibbs theory of capillarity is a thermodynamic system whose volume and shape are the same as in a real system and other properties (density, concentration of components, entropy and energy of the volume unit) constant up to the selected dividing surface.

Consequently, there is no heterogeneous fracture surface in the reference system between the phases, and the properties in transition from one phase to another change abruptly in the mathematical sense of the word. For determinacy, it must be accepted that the volume of the reference system remains equal to the volume of the real system even if the latter is deformed. In other words, the reference system should be selected to ensure that at any changes in the real system the ratio $V^\alpha + V^\beta = V$ holds, where V^α and V^β are the volumes of the phases in the reference system whose values depend on the position of the dividing surface, V is the volume of the real two-phase system whose value is independent of the position of the dividing surface.

Thus, the reference system in Gibbs capillarity theory has the form of two contacting volume phases 'superfluous' to the fracture surface. It should be mentioned that the phase equilibrium conditions were determined by Gibbs initially for these phases, *i.e.*, without taking into account the contribution of the fracture surface and, subsequently, it was shown that these conditions also hold if the contribution of the fracture surfaces is taken into account. It should also be mentioned that the condition of the quality of the volumes of the real thermodynamic system and the ideal reference system can be replaced, as shown in a series of investigations [21, 23, 28], by other conditions, if this simplifies the solution of the given tasks.

It is evident in no real system with the fracture surface can the properties of the phases remain constant up to the dividing surface, hypothetically made by the investigator in the region of heterogeneity. Therefore, the values of the extensive properties of the real system differ from the properties in the reference system. Gibbs quantitatively characterised these difference by excess quantities representing the differences between the appropriate extensive quantities in these systems. We examine this difference procedure in a greater detail because without clear understanding of the sense of the excess quantities it is not possible to understand and consequently use the Gibbs thermodynamic theory in practice.

Let us assume that a multiphase, multicomponent system with r

fracture surfaces between the phases is in the thermodynamic equilibrium condition.

The fundamental equation of thermodynamics for the multiphase system has the following form on the whole:

$$dU = TdS - PdV + \sum_{j=1}^{r} \sigma^j dA^j + \sum_{i=1}^{n} \mu_i dN_i \qquad (1.1.1)$$

where σ^j and A^j are the surface tension and the area of the j-th fracture surface, respectively. The quantities without the upper index relate to the entire system as a whole.

The useful property of the differential equation (1.1.1) is that it can be used equally for the entire system as a whole and also for any part of the system. Evidently, the sum $\sum_j^r \sigma^j dA^j$ contains quantities relating only to the surfaces situated inside the part for which the equation is written.

A part of the volume between the phases α and β is defined hypothetically in the form of a right-angled parallelepiped whose faces are normal to the fracture surface. The height of the parallelepiped is such that it contains the fracture surface and small interlayers of the volume phases V^α and V^β. The dividing surface A is produced inside the parallelepiped.

This procedure makes it possible to determine, without disrupting the strictness of the explanation, the boundaries for the investigated part of the system and avoid considering contributions brought in by other fracture surfaces.

In further considerations, the two phase real system will be represented by the part of the system enclosed in the specified parallelepiped. For the selected part containing one fracture surface, the fundamental equations of thermodynamics (1.1.1) gives:

$$dU = TdS - PdV + \sigma dA + \sum_{i=1}^{n} \mu_i dN_i \qquad (1.1.2)$$

The thermodynamic reference system is selected. According to the definition, in the reference system the properties of the phases α and β are constant up to the dividing surface and change abruptly on the surface. For homogeneous phases of the reference system:

$$dU^\alpha = TdS^\alpha - PdV^\alpha + \sum_{i=1}^{n} \mu_i dN_i^\alpha \qquad (1.1.3)$$

$$dU^\beta = TdS^\beta - PdV^\beta + \sum_{i=1}^{n} \mu_i dN_i^\beta \qquad (1.1.4)$$

The quantitative difference between the real system and the reference system is determined. The equations (1.1.3) and (1.1.4) are deducted from (1.1.2).

Taking into account $V = V - V^a - V^b = 0$ gives:

$$dU_\omega = TdS_\omega + \sigma dA + \sum_{i=1}^{n} \mu_i dN_{i\omega} \quad (1.1.5)$$

where $U_\omega = U - U^\alpha - U^\beta$, $S_\omega = S - S^\alpha - S^\beta$ and $N_{i\omega} = N_i - N_i^\alpha - Ni^\beta$ are excess quantities, characterising the difference between the real and reference system.

This equation shows that the excess internal energy may be regarded as a characteristic function of independent variables: excess entropy (S_ω), the area of the dividing surface (A), the excess number of particles ($N_{i\omega}$). Consequently, as in the case of the volume properties, the intensive properties may be expressed through the partial derivatives of the excess extensive properties, *i.e.*, Maxwell relationships also hold for excess properties:

$$\mathrm{T} = \left(\frac{\partial U_\omega}{\partial S_\omega}\right)_{A,\mathrm{N}_{i\omega}} ; \quad \sigma = \left(\frac{\partial U_\omega}{\partial A}\right)_{S_\omega,\mathrm{N}_{i\omega}} ; \quad \mu_i = \left(\frac{\partial U_\omega}{\partial \mathrm{N}_{i\omega}}\right)_{S_\omega,A,\mathrm{N}_{j\neq i}} \quad (1.1.6)$$

Since the internal energy of the phases and the excess internal energy are homogeneous functions of the first degree in relation to the independent variables (entropy, volume, number of particles), the Euler theorem gives:

$$U^\alpha = \mathrm{T}S^\alpha - \mathrm{P}V^\alpha + \sum_{i=1}^{n} \mu_i \mathrm{N}_i^\alpha \quad (1.1.7)$$

$$U^\beta = \mathrm{T}S^\beta - \mathrm{P}V^\beta + \sum_{i=1}^{n} \mu_i \mathrm{N}_i^\beta \quad (1.1.8)$$

$$U_\omega = \mathrm{T}S_\omega + \sigma A + \sum_{i=1}^{n} \mu_i \mathrm{N}_{i\omega} \quad (1.1.9)$$

Evidently, for a heterogeneous system containing r interfacial surfaces, the following equation is valid:

$$U = \mathrm{T}S - \mathrm{P}V + \sum_{j}^{r} \sigma^j A^j + \sum_{i}^{n} \mu_i \mathrm{N}_i \quad (1.1.10)$$

Since the intensive variables characterise the state of the system, integration at constant temperature, pressure, chemical potential and surface tension has the meaning of increase of the mass, volume and the area of the dividing surface without any change of its state.

Thus, the above relationships describe the internal energy of the volume phases α and β, the excess internal energy of the two-phase system with a single fracture surface, and a heterogeneous n-component multiphase system with r fracture surfaces.

1.2. Excess thermodynamic potential. Gibbs adsorption equation

As in the case of the volume phases, the fundamental equation for the excess quantities (1.1.10) may be converted and excess thermodynamic potentials of other variables may be introduced.

As an example, the excess free energy is examined:

$$F_\omega = U_\omega - S_\omega T = \sigma A + \sum_{i=1}^{n} \mu_i N_{i\omega} \tag{1.2.1}$$

Excess quantities depend on the position of the dividing surface. Therefore, selecting the dividing surface to fulfil the condition:

$$G_\omega = \sum_{i=1}^{n} \mu_i N_{i\omega} = 0 \tag{1.2.2}$$

(The fact that this selection of the dividing surface is justified will be shown later), equation (1.2.1) gives:

$$F_\omega = \sigma A \quad or \quad \frac{F_\omega}{A} = \sigma \tag{1.2.3}$$

where $\frac{F_\omega}{A}$ is the specific (related to the unit area) excess free energy, this parameter is referred to as the surface density of the free surface energy, G_ω is the Gibbs excess thermodynamic potential.

Equation (1.2.3) shows that surface tension is equal to the specific free surface energy only for a specific position of the dividing surface. For example, for a single-component system, equation (1.2.2) gives $\sigma = F_\omega / A$ only in the case in which the position of the dividing surface is determined by the condition $N_\omega = 0$. The surface, satisfying this condition, is referred to as the equimolar dividing surface. The dividing surface, satisfying the condition (1.2.2) is placed in a position in which the Gibbs excess thermodynamic potential is equal to zero. Therefore, it is referred to as the equienergetic dividing surface. For a one-component system, the positions of these dividing surfaces coincide.

It is believed erroneously [6, 29, 30] that equation (1.2.3) holds only for a single-component system in relation to the equimolar dividing surface, given by the condition $N_\omega = 0$.

However, the above considerations show that in selecting the dividing surface in accordance with the condition (1.2.2), the equality (1.2.3) is also fulfilled in multicomponent systems. Consequently, when the dividing surface is correctly selected, the surface tension may be equal numerically to the specific free surface energy also in multicomponent systems.

Surface tension is neither the internal nor potential energy of the unit area of the interfacial surface and is equal to the excess free surface energy of the unit surface for a specific position of the dividing surface.

Similarly, writing the equation for the differential of enthalpy and the Gibbs thermodynamic potential for the entire system as a whole and for the considered homogeneous phases α and β of the reference system, and carrying out the difference procedure, the following expressions are obtained:

a) for enthalpy

$$d\left(U+PV\right)=dH=TdS+VdP+\sigma dA+\sum_{i=1}^{n}\mu_i dN_i \qquad (1.2.4)$$

$$dH^{\alpha}=TdS^{\alpha}+V^{\alpha}dP+\sum_{i=1}^{n}\mu_i dN_i^{\alpha}$$

$$dH^{\beta}=TdS^{\beta}+V^{\beta}dP+\sum_{i=1}^{n}\mu_i dN_i^{\beta}$$

$$dH_{\omega}=TdS^{\alpha}+\sigma dA+\sum_{i=1}^{n}\mu_i dN_{i\omega} \qquad (1.2.5)$$

b) for the Gibbs thermodynamic potential:

$$dG=d\left(U-TS+PV\right)=-SdT+VdP+\sum_{i=1}^{n}\mu_i dN_I+\sigma dA \qquad (1.2.6)$$

$$dG^{\alpha}=-S^{\alpha}dT+V^{\alpha}dP+\sum_{i=1}^{n}\mu_i dN_i^{\alpha} \qquad (1.2.7)$$

$$dG^{\beta}=-S^{\beta}dT+V^{\beta}dP+\sum_{i=1}^{n}\mu_i dN_i^{\beta} \qquad (1.2.8)$$

$$dG_{\omega}=-S_{\omega}dT+\sigma dA+\sum_{i=1}^{n}\mu_i dN_{i\omega} \qquad (1.2.9)$$

Integrating the equations for the excess enthalpy and the Gibbs thermodynamic potential at constant temperature, surface tension and chemical potentials (T, σ, μ...) gives:

$$H_{\omega} = TS_{\omega} + \sigma A + \sum_{i=1}^{n} \mu_i N_{i\omega} = U_{\omega} \tag{1.2.10}$$

$$G_{\omega} = \sigma dA + \sum_{i=1}^{n} \mu_i N_{i\omega} = F_{\omega} \tag{1.2.11}$$

These equations show that in the case of excess quantities, excess enthalpy coincides with excess internal energy and the Gibbs potential with excess free energy. The result is evident because according to definition, these potentials differ from each other by the value PV, where P is pressure, V is volume, and in the case of excess quantities $V_{\omega} = V - V^{\alpha} - V^{\beta} = 0$, $PV_{\omega} = 0$.

For excess thermodynamic potentials one may obtain the equation which is an analogue of the Gibbs–Helmholtz equation. In fact, according to the definition of excess free energy:

$$F_{\omega} = U_{\omega} - TS_{\omega} = \sigma A$$

Consequently,

$$\sigma = \frac{U_{\omega}}{A} - T\frac{S_{\omega}}{A} = u_{\omega} - Ts_{\omega}$$

where u_{ω}, s_{ω} are the specific (related to the unit surface) excess internal energy and entropy.

As shown previously, $s_{\omega} = -\left(\frac{d\sigma}{dT}\right)_{A,N_{i\omega}}$ and, consequently: $u_{\omega} = \sigma - T\left(\frac{\partial\sigma}{\partial T}\right)_{A,N_{i\omega}}$

Surface tension is equal to the specific excess free energy only for the dividing surface which satisfies the condition (1.2.2). For a one-component system, this surface coincides with the equimolar surface. Therefore, the equation

$$u_{\omega} = \sigma - T\left(\frac{\partial\sigma}{\partial T}\right)_{A,N_{i\omega}} \tag{1.2.12}$$

holds only for this dividing surface.

The dividing surface satisfying the condition (1.2.2), has a specific physical meaning. Since it is generated in such a manner that the Gibbs excess thermodynamic potential is equal to zero, this dividing surface may be referred to as the equienergetic dividing surface.

Open systems may be described efficiently using a high or total thermodynamic potential. For a two-phase, multicomponent system:

$$\Omega = U - TS - \sum_{i=1}^{n} \mu_i N_i = \sigma A - PV \tag{1.2.13}$$

$$d\Omega = -SdT - PdV + \sigma dA - \sum_{i=1}^{n} \mathrm{N}_i d\mu_i \tag{1.2.14}$$

Therefore,

$$S = -\left(\frac{\partial\Omega}{\partial \mathrm{T}}\right)_{V,A,\mu_i}; P = -\left(\frac{\partial\Omega}{\partial V}\right)_{V,T,A}; \tag{1.2.15}$$

$$\sigma = \left(\frac{\partial\Omega}{\partial A}\right)_{T,\mu,V}; \ \mathrm{N}_i = -\left(\frac{\partial\Omega}{\partial \mu_i}\right)_{V,T,A} \tag{1.2.16}$$

These equations may be used to determine the relationship between excess free energy and the excess high thermodynamic potential. In fact:

$$\Omega_\omega = F_\omega - \sum_{i=1}^{n} \mu_i \mathrm{N}_{i\omega} = U_\omega - TS_\omega - \sum_{i=1}^{n} \mu_i \mathrm{N}_{i\omega} = \sigma A \tag{1.2.17}$$

From the differential of the high thermodynamic potential:

$$d\Omega_\omega = -S_\omega d\mathrm{T} - \sigma dA - \sum_{i=1}^{n} \mathrm{N}_{i\omega} d\mu_i \tag{1.2.18}$$

This gives:

$$S_\omega = -\left(\frac{\partial\Omega_\omega}{\partial T}\right)_{A,\mu_i}; \sigma = -\left(\frac{\partial\Omega_\omega}{\partial A}\right)_{T,\mu}; \mathrm{N}_{i\omega} = -\left(\frac{\partial\Omega_\omega}{\partial \mu_i}\right)_{T,A,\mu_{j\neq i}} \tag{1.2.19}$$

Using the resultant expressions, it is easy to determine the relationship between F_ω and Ω_ω. Since

$$\Omega_\omega = F_\omega - \sum_{i=1}^{n} \mu_i \mathrm{N}_{i\omega} \tag{1.2.20}$$

$$\mathrm{N}_{i\omega} = -\left(\frac{\partial\Omega_\omega}{\partial \mu_i}\right)_{T,a,\mu_{j\neq i}} \tag{1.2.21}$$

then

$$F_\omega = \Omega_\omega - \sum_{i=1}^{n} \mu_i \left(\frac{\partial\Omega_\omega}{\partial \mu_i}\right)_{T,A,\mu_{j\neq i}} \tag{1.2.22}$$

Thus, like the thermodynamic potentials of the volume phases, the excess thermodynamic potentials are linked together by fundamental equations and, therefore, they may be examined as if they characterised some hypothetical phase.

In contrast to the thermodynamic potentials of the volume phases, the excess quantities depend on the position of the dividing surface. An exception is the high thermodynamic potential which, as shown later, is invariant in relation to the hypothetical displacement of the dividing surface. However, regardless of the dependence of the excess thermodynamic potentials on the position of the dividing surface, the general form of the thermodynamic equations, with the exception of special cases, is invariant in relation to the shift of the dividing surface.

The intensive properties of the volume phases are linked together by the Gibbs–Duhem equation which will be examined in detail later. An identical equation was derived by Gibbs also for excess thermodynamic parameters of the interfacial layer. The following equations were derived previously for the excess high thermodynamic potential:

$$\Omega_{\omega} = \sigma A \tag{1.2.23}$$

$$d\Omega_{\omega} = -S_{\omega} d\mathrm{T} + \sigma d\mathrm{A} - \sum_{i=1}^{n} N_{i\omega} d\mu_i \tag{1.2.24}$$

Equation (1.2.3.1.2) is differentiated and the resultant equation is compared with (1.2.3.2.1). This gives an equation which is the analogue of the Gibbs–Duhem equation and plays a significant role in the Gibbs theory of capillarity:

$$S_{\omega} d\mathrm{T} + A d\sigma + \sum_{i=1}^{n} N_{i\omega} d\mu_i = 0 \tag{1.2.25}$$

Taking into account that the quantities S_{ω} and $N_{i\omega}$ are proportional to the area of the dividing surface A, the resultant equation is divided by A. This equation is referred to as the Gibbs adsorption equation and has the following form:

$$d\sigma = -S_{\omega} d\mathrm{T} - \sum_{i=0}^{n} \Gamma_i d\mu_i \tag{1.2.26}$$

For convenience, the equation (1.2.26) contains the following notations:

$$\Gamma_i = \frac{N_i - N_i^{\alpha} - N_i^{\beta}}{A} \quad S_{\omega} = \frac{S - S^{\alpha} - S^{\beta}}{A} \tag{1.2.27}$$

where S_{ω}, Γ_i is the excess entropy and the excess number of particles per unit area of the dividing surface. The excess number of the particles of the i-th component Γ_i is referred to as the adsorption of this component, and equation (1.2.26) as the Gibbs adsorption equation.

The physical quantities without the indexes (N_i, S) characterise the two-phase system as a whole and do not depend on the position of the dividing surface. The quantities, denoted by the indexes α and β ($N_i^\alpha, N_i^\beta, S^\alpha, S^\beta$) relate to the discussed phases α and β of the reference system. This means that these are the values of the properties which the phases α and β would have if the density of these properties remained constant up to the selected dividing surface. Evidently, the values of these quantities depend on the position of the dividing surface.

As indicated by the above considerations, the values of the extensive properties, included in the Gibbs adsorption equation, depend on the position of the dividing surface. Therefore, it would be more accurate to refer to them as the excess quantities in relation to the given dividing surface. However, to simplify considerations, Gibbs related these quantities to the dividing surface and referred to them as the surface properties, assuming that 'this would not cause any misunderstanding' [15]. Thus, U_ω and S_ω are referred to as surface energy and surface entropy, and Γ_i is the adsorption of the component on the given surface.

Therefore, the special feature of the excess thermodynamic quantities is that they depend on the position of the hypothetically selected dividing surface. In order to apply the Gibbs theory efficiently, it is necessary to explain the nature of this dependence. This problem is not reflected in the scientific literature, although a deeper understanding of the Gibbs theory of capillarity is not possible without obtaining clear information on the excess properties and the dependence of these properties on the position of the dividing surface. Therefore, this problem will be examined in greater detail.

1.3. Dependence of excess thermodynamic quantities on the position of the dividing surface

One of the successes by Gibbs was that he paid attention for the first time to the importance of the extensive parameters (in addition to the intensive parameters) when describing a thermodynamic system. As shown in [15], if the corresponding densities in the phases are introduced for the extensive quantities, their role is as important as that of the intensive quantities. This can be clearly seen when examining the dependence of the excess thermodynamic quantities of the position of the dividing surface.

Any excess extensive thermodynamic quantity may be represented in the form:

$$X_{\omega} = \frac{X - X^{\alpha} - X^{\beta}}{A} \tag{1.3.1}$$

where X is the value of the property in a real two-phase system; X^{α} and X^{β} the values of these properties which the phases α and β would have if the density of this property in the phase α would remain constant up to the dividing surface.

If we introduce the concept of the local property, then any extensive property of the two-phase system in transition through the fracture surface is characterised by its profile $x(z)$. Evidently, for any extensive property of the two-phase system the following equation is valid:

$$X = A\int_{z^{\alpha}}^{z^{\beta}} x(z)dz \tag{1.3.2}$$

where z^{α} and z^{β} are the coordinates of the phase boundary; X is the value of the property in the two-phase system, $x(z)$ is the function which gives the dependence of the local density of the investigated property on the normal to the surface of the coordinate.

Similarly, we can present the value of the property in the idealised reference system. Consequently, for the excess thermodynamic quantity [8]:

$$X_{\omega}^{0} = \int_{z^{\alpha}}^{0}\left[x(z) - x^{\alpha}\right]dz + \int_{0}^{z^{\beta}}\left[x(z) - x^{\beta}\right]dz \tag{1.3.3}$$

Equation (1.3.3) can be used to determine the dependence of the excess properties on the position of the dividing surface. The position of the dividing surface is determined by the normal coordinate z. Equation (1.3.2.1) is written for the case in which the dividing surface is given by the coordinate z = 0. The surface is moved hypothetically over the distance Δz in the direction of the β-phase (the z axis is directed from the α-phase to the β-phase). Consequently, for the excess property in relation to the new position of the dividing surface, the equation (1.3.3) is written in the new form:

$$X_{\omega} = \int_{z^{\alpha}}^{\Delta z}\left[x(z) - x^{\alpha}\right]dz + \int_{\Delta z}^{z^{\beta}}\left[x(z) - x^{\beta}\right]dz \tag{1.3.4}$$

The resultant equation may be easily presented in the form:

$$X_\omega = \int_{z^\alpha}^{0}\left[x(z)-x^\alpha\right]dz + \int_{0}^{z^\beta}\left[x(z)-x^\beta\right]dz + \int_{0}^{\Delta z}\left[x(z)-x^\alpha-\left(x(z)-x^\beta\right)\right]dz \tag{1.3.5}$$

From (1.3.5), taking equation (1.3.3) into account, we obtain:

$$X_\omega = X_\omega^0 - \int_0^{\Delta z}\left(x^\alpha - x^\beta\right)dz$$

or

$$X_\omega = X_\omega^0 - \left(x^\alpha - x^\beta\right)\Delta z \tag{1.3.6}$$

where x^α and x^β is the density of the properties in the phases α and β; Δz the distance over which the dividing surface has been displaced.

Consequently, for the appropriate excess properties:

$$\left.\begin{aligned}
U_\omega &= U_\omega^0 - \left(u^\alpha - u^\beta\right)z;\\
S_\omega &= S_\omega^0 - \left(s^\alpha - s^\beta\right)z;\\
F_\omega &= F_\omega^0 - \left(f^\alpha - f^\beta\right)z;\\
\Gamma_i &= \Gamma_i^0 - \left(n_i^\alpha - n_i^\beta\right)z;\\
g_\omega &= g_\omega^0 - \left(g_i^\alpha - g_i^\beta\right)z;
\end{aligned}\right\} \tag{1.3.7}$$

The resultant equations show that the dependence of the excess quantities on the normal to the surface of the coordinate z is linear. Consequently, it is possible to 'fix' the dividing surface, in relation to the given problem and in relation to any excess property.

If the values of adsorption on different dividing surfaces are available, equations (1.3.7) may be used for determining the distance between these surfaces:

$$Z = -\frac{\Delta\Gamma_i}{n_i^\alpha - n_i^\beta} \tag{1.3.8}$$

Thus, for the distance between the Gibbs Γ_i and the equimolar dividing surfaces $\sum_{i=1}^{n}\Gamma_{i(n)} = 0$:

$$Z_{i-N} = -\frac{\Gamma_{i(n)}}{n_i^\alpha - n_i^\beta} \tag{1.3.9}$$

Calculations carried out using equation (1.3.8) show [31] that the

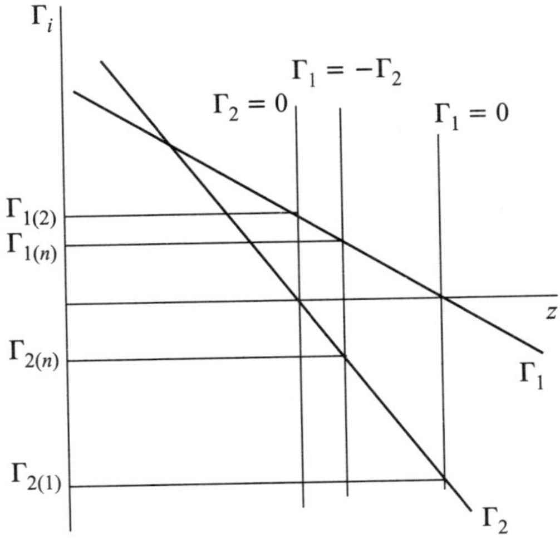

Figure 1.1. Various positions of the dividing surface used in the calculations.

distance greatly changes with the variation of the composition of the melt. It may easily be shown that the Gibbs dividing surface moves away from the interphase boundary.

According to (1.3.7), the values of the excess quantities, including adsorption, depend in a linear manner on the position of the dividing surface. For better understanding, Fig. 1.1 gives the graphs of the dependence of the adsorption of components in a two-component system. The same figure shows different positions on the dividing surface ($\Gamma_i = 0$, $-\Gamma_1 = \Gamma_2$).

From the viewpoint of mathematics, all the position of the dividing surface are equally important since the Gibbs adsorption equation holds for any position of the surface. However, as noted by Gibbs, from the physical viewpoint it is more convenient to place the dividing surface in the region of heterogeneity, *i.e.* inside the fracture surface. The correctness of this selection is also indicated by the fact that the surface tension is attributed to this dividing surface.

Gibbs proposed to calculate the adsorption of components in relation to the dividing surface, generated in such a manner that the adsorption of one of the components was equal to zero. The adsorption of the components in relation to the surface is referred to as the Gibbs relative adsorption and denoted by $\Gamma_{i(j)}$. Index j denotes that the adsorption of the j-th component on the surface is equal to zero. For a single-component system, this surface coincides with the equimolar surface.

Later, analysing the adsorption phenomena in the liquid–vapour system, Guggenheim and Adam proposed methods of selecting the dividing surface. As indicated by Fig. 1.1, in the vicinity of the fracture surface there is, regardless of the type of density profile, a position of the dividing surface $\sum_{i=1}^{n}\Gamma_i=0$ in which the condition is fulfilled. In the two-component system this condition has the form $\Gamma_1 = -\Gamma_2$.

In the literature, the adsorption in relation to the dividing surface is referred to as the N-variant of adsorption according to Guggenheim and Adam. However, it is more convenient to refer to the surface by analogy with the single-component system as the equimolar dividing surface, and the adsorption in relation to the surface as the adsorption on the equimolar dividing surface. Guggenheim and Adam also proposed that the methods of determination of adsorption which, as shown later, differ from the examined variants only by the method of selecting the dividing surface [32–34]. In particular, it was proposed to calculate the adsorption in relation to the dividing surfaces, corresponding to the conditions $\sum_{i=0}^{n}\Gamma_i M_i=0$ and $\sum_{i=0}^{n}\Gamma_i \bar{V}_i=0$, where M_i is the molar mass of the component, $\bar{V}_i$ is the partial molar volume of the i-th component in one of the phases.

Thus, depending on the investigated problem, the dividing surface, may be positioned in such a manner that one of the following conditions is satisfied:

$$U_\omega=0;\ S_\omega=0\ \sum_{i=1}^{n}\Gamma_i\mu_i=0;\ \Gamma_J=0;\ \sum_{i=1}^{n}\Gamma_{i(n)}=0\ \sum_{i=1}^{n}\Gamma_i\bar{V}_i=0;\ \sum_{i=1}^{n}\Gamma_i\bar{V}_i=0 \tag{1.3.10}$$

In most cases, adsorption is calculated using the last four conditions. To simplify calculations, some of the conditions (1.3.10) may be generalised, as shown in [22], by the expression:

$$\sum_{i=1}^{n}\Gamma_{i(p)}P_i=0 \tag{1.3.11}$$

where P_i depends on the selection of the dividing surface. For example, for adsorption on the equimolar dividing surface $P_i = 1$ and $\sum_{i=1}^{n}\Gamma_{i(n)}=0$, for the M-variant of adsorption $P_i = M_i$ and $\sum_{i=1}^{n}M_i\Gamma_{i(M)}=0$, and for adsorption on the physical dividing surface $P_i=\bar{V}_i$ and $\sum_{i=1}^{n}\Gamma_{i(V)}\bar{V}_i=0$.

In scientific literature, special attention is often given to the problem of advantages or shortcomings of a specific method of determination of adsorption. For example, according to the authors of [27], the conditions $\Gamma = 0$ may be fulfilled only in a diluted solutions and, consequently, it is not correct to calculate the relative Gibbs adsorption in the entire range of the compositions. Other authors [268] assumed that it is not correct to calculate Gibbs adsorption in concentration solutions since Gibbs adsorption does not tend to zero $\Gamma_2 \neq 0$ at $x_2 = 0$. The error of these considerations is evident from the previously discussed matters.

The authors of [35] assumed that the dividing surface whose position does not change with composition should be preferred and they used the V-variant.

On the other hand, in the determination of the mean composition of the surface layer it is more convenient to calculate the adsorption on the equimolar dividing surface (N-variant according to Guggenheim and Adam) since the value of this adsorption is linked by a simple relationship with the average composition of the surface layer.

As shown previously, in the thermodynamic respect, all the methods of determination of adsorption are equivalent and differ from each other only by the method of 'fixation' of the dividing surface. Therefore, the selection of the position of the dividing surface should be determined by the ease of solving the given problem [27, 28]. In [41, 43–45] it is shown that the general form of the equation for adsorption excesses may be determined without using the concept of the dividing surface.

We examine in greater detail the problem of calculating the adsorption in relation to different positions of the dividing surface in multicomponent systems on the basis of the concentration dependence of surface tension and chemical potential. For a two-component system this problem is quite easy to solve. For this purpose it is sufficient to supplement the Gibbs adsorption equation by the condition of fixation of the position of the dividing surface. Therefore, the problem of calculating adsorption in binary systems will be investigated in the initial stage.

The increase of the number of components also increases the number of unknown quantities in the Gibbs adsorption equation and the problem of calculating adsorption is slightly more complicated.

1.4. Adsorption equilibrium in two-phase multicomponent systems

The Gibbs adsorption equation will be supplemented by one of the conditions fixing the position of the dividing surface. Consequently, for the relative Gibbs adsorption $\Gamma_j = 0$ we obtain:

$$-d\sigma = -S_{\omega(j)}dT - \sum_{i=1}^{n}\Gamma_{i(j)}\left(1-\delta_{ij}\right)d\mu_i \quad (1.4.1)$$

where $\delta_{ij} = \begin{cases} 0, \text{ at } i \neq j \\ 1, \text{ at } i = j \end{cases}$ is the Kronecker symbol, and the lower index j indicates that it is the value of the excess property in relation to the dividing surface, corresponding to the zero value of adsorption of the j-th component.

In equation (1.4.1) all the excess quantities are related to the specific dividing surface. As shown later, the equation may be used to calculate the relative adsorption of components $\Gamma_{i(j)}$, if the concentration dependence of surface tension and chemical potentials is known.

Similarly, we can derive an equation for calculating the adsorption in relation to other dividing surfaces. However, for this purpose it is convenient to transform the Gibbs adsorption equation taking into account the phase equilibria conditions. If the phases α and β, separated by the flat fracture surface, are situated in the thermodynamic equilibrium condition, then for the infinitely small equilibria change we can write the following expressions [19]:

$$\begin{aligned} dP^{\alpha} &= dP^{\beta}, \\ d\mu_i^{\alpha} &= d\mu_i^{\beta}, \\ dT^{\alpha} &= dT^{\beta} \end{aligned} \quad (1.4.2)$$

On the other hand, for every of the coexisting phases:

$$dP^{\alpha} = s^{\alpha}dT^{\alpha} + \sum_{i=1}^{n} n_i^{\alpha} d\mu_i^{\alpha} \quad (1.4.3)$$

$$dP^{\beta} = s^{\beta}dT^{\beta} + \sum_{i=1}^{n} n_i^{\beta} d\mu_i^{\beta} \quad (1.4.4)$$

Taking into account (1.4.2), from (1.4.3) and (1.4.4) we obtain:

$$\left(s^{\alpha} - s^{\beta}\right)dT + \sum_{i=1}^{n}\left(n_i^{\alpha} - n_i^{\beta}\right)d\mu_i = 0 \quad (1.4.5)$$

where $s^{\alpha} = \frac{S^{\alpha}}{V^{\alpha}}$ and $s^{\beta} = \frac{S^{\beta}}{V^{\beta}}$ are the densities of entropy in the phases α and β, n_i^{α} and n_i^{β} are the densities of the number of particles of the i-th type in the phases α and β.

Equation (1.4.5) is the Gibbs–Duhem equation for the equilibrium two-phase system. Evidently, for the isothermal processes ($dT = 0$) this equation has the following form:

$$\sum_{i=1}^{n}\left(n_i^\alpha - n_i^\beta\right)d\mu_i = 0 \tag{1.4.6}$$

In the literature, equation (1.4.6) is often replaced, without additional explanation, by the Gibbs–Duhem equation in the form:

$$\sum_{i=1}^{n} x_i^\alpha d\mu_i = 0, \tag{1.4.7}$$

where x_i^α is the molar fraction of the component in the α-phase.

It may easily be shown that (1.4.6) transforms to (1.4.7) only when the content of matter in one of the phases may be ignored, i.e. the condition $\left(\sum_{i=1}^{n} n_i^\beta \ll \sum_{i=1}^{n} n_i^\alpha\right)$ is fulfilled.

Thus, the equation (1.4.7) may be used only in cases in which the content of the component in one of the phases may be ignored. However, if the number of components is greater than two, an additional degree of freedom appears in the system and the composition of the α-phase may change along isotherms–isobars in such a manner that the pressure and temperature of the two phase systems remain constant. For this method of changing the composition, equation (1.4.7) is derived directly from (1.4.4).

Thus, for the equilibrium two phase system we can write:

$$d\sigma = -S_\omega dT - \sum_{i=0}^{n}\Gamma_i d\mu_i \tag{1.4.8}$$

$$\left(s^\alpha - s^\beta\right)dT + \sum_{i=1}^{n}\left(n_i^\alpha - n_i^\beta\right)d\mu_i = 0 \tag{1.4.9}$$

Expressing $d\mu_j$ from equation (1.4.9) and substituting into the Gibbs adsorption equation (1.2.4) gives:

$$-d\sigma = \left(S_\omega - \Gamma_j \frac{s^\alpha - s^\beta}{n_j^\alpha - n_j^\beta}\right)dT + \sum_{\substack{i=1 \\ i\neq j}}^{n}\left[\Gamma_i - \Gamma_j \frac{n_i^\alpha - n_i^\beta}{n_j^\alpha - n_j^\beta}\right]d\mu_i \tag{1.4.10}$$

In the resultant expression, $d\sigma$ is the total differential. Consequently, the coefficient at dT and $d\mu_i$ in this equation are partial derivatives of surface tension in respect of temperature and chemical potentials:

$$S_\omega - \Gamma_j \frac{S^\alpha - S^\beta}{n_j^\alpha - n_j^\beta} = -\left(\frac{\partial\sigma}{\partial T}\right)_{\mu_{i\neq j}} \tag{1.4.11}$$

$$\Gamma_i - \frac{\Gamma_j}{n_j^\alpha - n_j^\beta}\left(n_i^\alpha - n_i^\beta\right) = \left(\frac{\partial\sigma}{\partial\mu_i}\right)_{T,\mu_{K\neq i-j}} \tag{1.4.12}$$

It may easily be shown that the numerical value of the coefficients at dT and $d\mu_i$ in equation (1.27) coincides with the values of the appropriate quantities in relation to the dividing surface, satisfying the condition $\Gamma_j = 0$.

In fact, $\frac{\Gamma_j}{n_j^\alpha - n_j^\beta} = Z_{j(0)}$ is the distance from the selected dividing surface of the surface given by the condition $\Gamma_j = 0$. Consequently, from (1.4.12) we obtain:

$$\Gamma_i - \frac{\Gamma_j}{n_j^\alpha - n_j^\beta}\left(n_i^\alpha - n_i^\beta\right) = \Gamma_i - Z_{j(0)}\left(n_i^\alpha - n_i^\beta\right) = \Gamma_{i(j)} \tag{1.4.13}$$

Similarly

$$S_\omega - \Gamma_j \frac{s^\alpha - s^\beta}{n_j^\alpha - n_j^\beta} = S_\omega - Z_{j(0)}\left(s^\alpha - s^\beta\right) = S_{\omega(j)} \tag{1.4.14}$$

Taking into account (1.4.13) and (1.4.14), the equations (1.4.11) and (1.4.12) may be presented in a new form:

$$S_{\omega(j)} = \left(\frac{\partial\sigma}{\partial \mathrm{T}}\right)_{\mu_{i\neq j}} \quad \text{and} \quad \Gamma_{i(j)} = \left(\frac{\partial\sigma}{\partial\mu_i}\right)_{\mathrm{T},\mu_{k\neq j\neq i}}$$

Consequently, the selection of the equimolecular dividing surface in relation to some component ($\Gamma_j = 0$) in the description of the adsorption processes is equivalent to taking into account the equilibrium conditions of the fracture surface and the coexisting phases.

The methods of calculating adsorption, proposed by Guggenheim and Adam, will be investigated. For this purpose, it is necessary to supplement the equation (1.4.10) by the relationship, determining the link between the adsorption values. It is written in the general form:

$$\sum_{k=1}^{n} P_k \Gamma_{k(P)} = 0 \tag{1.4.15}$$

Taking into account (1.4.13) and (1.4.15), the equation (1.4.12) is written in the new form:

$$\Gamma_{i(P)} + \sum_{\substack{k=1\\k\neq j}}^{n} \frac{P_k \Gamma_{k(P)}}{P_j} \frac{n_i^\alpha - n_i^\beta}{n_j^\alpha - n_j^\beta} = -\left(\frac{\partial\sigma}{\partial\mu_i}\right)_{T,\mu_{k+j-i}} \tag{1.4.16}$$

The index (P) shows that this is the adsorption on the dividing surface, given by the condition (1.4.15).

If the system contains n components, then i in (1.4.16) acquires $n-1$ values 1, 2,..., $j-1$, $j+1$,..., n, and this expression represents a system of linear equations. Solving the resultant system in relation to $\Gamma_i(P)$, it is possible to determine their values or expressed these values through $\Gamma_{i(j)}$, *i.e.*, knowing the values of the relative adsorption of the component, we can determine adsorption on other dividing surfaces. It is believed that it is interesting to examine this problem in greater detail and this will carried out later. To provide more information and explain the sequence of explanation, it is appears to be useful to examine in advance the case of a two-phase, two-component system.

1.5. Calculation of adsorption in two-component systems

The calculation of adsorption of components on the basis of the concentration dependence of the surface tension is greatly facilitated in the case of two-component systems. In this case, the Gibbs adsorption equation gives:

$$-d\sigma = \left(S_\omega - \Gamma_2 \frac{s^\alpha - s^\beta}{n_2^\alpha - n_2^\beta} \right) dT + \left(\Gamma_1 - \Gamma_2 \frac{n_1^\alpha - n_1^\beta}{n_2^\alpha - n_2^\beta} \right) d\mu_1 \quad (1.5.1)$$

or at $\Gamma_2 = 0$

$$-d\sigma = S_{\omega(2)} d\mathrm{T} + \Gamma_{1(2)} d\mu_1 \quad (1.5.2)$$

Equation (1.5.2) shows that:

$$S_{\omega(2)} = -\left(\frac{\partial \sigma}{\partial \mathrm{T}} \right)_{\mu_1} \quad (1.5.3)$$

$$\Gamma_{1(2)} = -\left(\frac{\partial \sigma}{\partial \mu_1} \right)_{\mathrm{T}} \quad (1.5.4)$$

If the dependence of surface tension on chemical potential and temperature is determined by experiments or theoretically, then (1.5.3) and (1.5.4) may be used to determine the density of excess entropy and the number of particles or, according to the introduced terminology, surface entropy and relative Gibbs adsorption.

As mentioned previously, adsorption may be determined in relation to other positions of the dividing surface. Actually, if the position of the dividing surface satisfies the condition $\Gamma_1 P_1 - \Gamma_2 P_2 = 0$ equation (1.5.1) gives:

$$-d\sigma = \left(S_{\omega(P)} - \Gamma_{1(P)} \frac{P_1}{P_2} \frac{s_\alpha - s_\beta}{n_2^\alpha - n_2^\beta} \right) dT + \left(\Gamma_{1(P)} + \Gamma_{1(P)} \frac{P_1}{P_2} \frac{n_1^\alpha - n_1^\beta}{n_2^\alpha - n_2^\beta} \right) d\mu_1 \quad (1.5.5)$$

Equation (1.5.5) gives for adsorption $\Gamma_{i(P)}$:

$$\Gamma_{1(P)} = -\frac{1}{\left(1 + \frac{P_1}{P_2} \frac{n_1^\alpha - n_1^\beta}{n_2^\alpha - n_2^\beta}\right)} \left(\frac{\partial \sigma}{\partial \mu_1}\right)_T = -\frac{P_2\left(n_2^\alpha - n_2^\beta\right)}{P^\alpha - P^\beta} \left(\frac{\partial \sigma}{\partial \mu_1}\right)_T \quad (1.5.6)$$

The differential equation (1.5.5) may be used to determine the expression for excess entropy $S_{\omega(P)}$:

$$S_{\omega(P)} = -\left(\frac{\partial \sigma}{\partial T}\right)_{\mu_1} - \Gamma_{1(P)} \frac{P_1\left(s^\alpha - s^\beta\right)}{P_2\left(n_2^\alpha - n_2^\beta\right)} \quad (1.5.7)$$

Or, taking into account the expressions (1.5.6), (1.5.4), (1.5.3), the resultant expression may be transformed to:

$$S_{\omega(P)} = -\left(\frac{\partial \sigma}{\partial T}\right)_{\mu_1} + \frac{P_1\left(s^\alpha - s^\beta\right)}{P^\alpha - P^\beta} \left(\frac{\partial \sigma}{\partial \mu_1}\right)_T = S_{\omega(2)} - \frac{P_1\left(s^\alpha - s^\beta\right)}{P^\alpha - P^\beta} \Gamma_{1(2)} \quad (1.5.8)$$

where

$$P^\alpha = P_2 n_2^\alpha + P_1 n_1^\alpha; \qquad P^\beta = P_2 n_2^\beta + P_1 n_1^\beta \quad (1.5.9)$$

The independent variable in the resultant equations is represented by the chemical potential which is not always convenient because the chemical potential is not determined directly in experiments. Therefore, we transfer from the chemical potentials to molar fractions and thermodynamic activity coefficient. For this purpose, we use the well-known equation for the chemical potential:

$$\mu_i = \mu_{i0} + RT \ln a_i = \mu_{i0} + RT \ln f_i x_i \quad (1.15.10)$$

where $a_i = f_i x_i$ is the thermodynamic activity, f_i is the thermodynamic activity coefficient. It is evident that at T = const

$$d\mu_i = RTd \ln a_i = RT \frac{da_1}{a_1} \quad (1.5.11)$$

$$da_1 = \partial\left(f_1, x_1\right) = x_1 \frac{\partial f_1}{\partial x_1} dx_1 + f_1 dx_1 = \left(x_1 \frac{\partial f_1}{\partial x_1} + f_1 \right) dx_1 \quad (1.5.12)$$

Taking into account (1.5.12), the resultant equations may be written in the new form:

$$\Gamma_{1(2)} = -\left(\frac{\partial\sigma}{\partial\mu_1}\right)_T = -\frac{a_1}{RT}\left(\frac{\partial\sigma}{\partial a_1}\right) = -\frac{f_1 x_1}{RT}\frac{\partial\sigma}{\partial(f_1;x_1)} \tag{1.5.13}$$

or

$$\Gamma_{1(2)} = -\frac{f_1 x_1}{RT\left(x_1\dfrac{\partial f_1}{\partial x_1} + f_1\right)}\left(\frac{\partial\sigma}{\partial x_1}\right)_T \tag{1.5.14}$$

$$S_{\omega(2)} = -\left(\frac{\partial\sigma}{\partial T}\right)_{a_1} \tag{1.5.15}$$

$$\Gamma_{1(P)} = -\frac{P_2\left(n_2^\alpha - n_2^\beta\right)a_1}{\left(P^\alpha - P^\beta\right)RT}\left(\frac{\partial\sigma}{\partial a_1}\right)_T \tag{1.5.16}$$

$$S_{\omega(P)} = -\left(\frac{\partial\sigma}{\partial T}\right)_{\mu_1} + \frac{P_1 a_1\left(s^\alpha - s^\beta\right)}{RT\left(P^\alpha - P^\beta\right)}\left(\frac{\partial\sigma}{\partial a_1}\right)_T \tag{1.5.17}$$

$$S_{\omega(P)} = S_{\omega(2)} - \frac{P_1}{P^\alpha - P^\beta}\Gamma_{1(2)}\left(s^\alpha - s^\beta\right) \tag{1.5.18}$$

Thus, the resultant expressions may be used to calculate the adsorption and excess entropy in relation to different positions of the dividing surface with the variation of surface tension with composition and temperature. The last equation (1.5.18) establishes the relationship between the value of the excess entropy in relation to different positions of the dividing surface.

The equations (1.5.16)–(1.5.18) hold for specific positions of the dividing surface determined by the parameter P_i. For example, for the equimolar dividing surface $P_i = 1$, and from (1.5.9)–(1.5.18) we have:

$$P^\alpha = P_2 n_2^\alpha + P_1 n_1^\alpha = n_2^\alpha + n_1^\alpha = n^\alpha$$

$$P^\beta = P_2 n_2^\beta + P_1 n_1^\beta = n_2^\beta + n_1^\beta = n^\beta$$

$$\Gamma_{1(n)} = -\frac{\left(n_2^\alpha - n_2^\beta\right)a_1}{\left(n^\alpha - n^\beta\right)RT}\left(\frac{\partial\sigma}{\partial a_1}\right)_T \tag{1.5.19}$$

$$S_{\omega(n)} = -\left(\frac{\partial\sigma}{\partial T}\right)_{\mu_1} + \frac{a_1\left(s^\alpha - s^\beta\right)}{RT\left(n^\alpha - n^\beta\right)}\left(\frac{\partial\sigma}{\partial a_1}\right) \tag{1.5.20}$$

$$S_{\omega(n)} = S_{\omega(2)} - \frac{s^\alpha - s^\beta}{n^\alpha - n^\beta}\Gamma_{1(2)} \tag{1.5.21}$$

Similarly, for the equimass dividing surface $P_i = M_i$

$$P^\alpha = M_2 n_2^\alpha + M_1 n_1^\alpha = \rho_2^\alpha + \rho_1^\alpha = \rho^\alpha \tag{1.5.22}$$

$$P^\beta = M_2 n_2^\beta + M_1 n_1^\beta = \rho_2^\beta + \rho_1^\beta = \rho^\beta \tag{1.5.23}$$

$$\Gamma_{1(M)} = -\frac{\left(\rho_2^\alpha - \rho_2^\beta\right)a_1}{\left(\rho^\alpha - \rho^\beta\right)RT}\left(\frac{\partial\sigma}{\partial a_1}\right)_T \tag{1.5.24}$$

where ρ^α and ρ^β are the densities of the phases α and β, ρ_2^α and ρ_2^β is the mass of the second component of the unit volume of the phases α and β.

For surface entropy:

$$S_{\omega(n)} = -\left(\frac{\partial\sigma}{\partial T}\right)_{\mu_1} + \frac{M_1 a_1\left(s^\alpha - s^\beta\right)}{RT\left(\rho^\alpha - \rho^\beta\right)}\left(\frac{\partial\sigma}{\partial a_1}\right)_T \tag{1.5.25}$$

$$S_{\omega(M)} = S_{\omega(2)} - \frac{M_1\left(s^\beta - s^\alpha\right)}{\rho^\alpha - \rho^\beta}\Gamma_{1(2)} \tag{1.5.26}$$

For a physical dividing surface $P_i = \overline{V}_i^\alpha$:

$P^\alpha = \overline{V}_2^\alpha n_2^\alpha + \overline{V}_1^\alpha n_1^\alpha = V^\alpha$ V^a – the molar volume of the α-phase,

$P^\beta = \overline{V}_2^\beta n_2^\beta + \overline{V}_1^\beta n_1^\beta = V^{\beta\alpha}$ V^{ab} – the hypothetical molar volume, *i.e.*, the volume in the β-phase which would exist if the partial–molar volumes of the phases α and β would be identical and equal to $\overline{V}_i^\alpha$:

$$\Gamma_{i(V)} = -\frac{\overline{V}_2\left(n_1^\alpha - n_2^\beta\right)a_1}{(V^\alpha - V^{\beta\alpha})RT}\left(\frac{\partial\sigma}{\partial a_1}\right)_T \tag{1.5.27}$$

$$\Gamma_{1(V)} = -\frac{\left(v_2^\alpha - v_2^\beta\right)a_1}{(V^\alpha - V^{\beta\alpha})RT}\left(\frac{\partial\sigma}{\partial a_1}\right)_T \tag{1.5.28}$$

$$S_{\omega(V)} = -\left(\frac{\partial\sigma}{\partial T}\right)_{\mu_1} + \frac{\overline{V}_1^\alpha a_1\left(s^\alpha - s^\beta\right)}{RT\left(V^\alpha - V^{\beta\alpha}\right)}\left(\frac{\partial\sigma}{\partial a_1}\right)_T$$

$$S_{\omega(V)} = S_{\omega(2)} - \frac{\overline{V}_1^{\alpha}\left(s^{\alpha} - s^{\beta}\right)}{V^{\alpha} - V^{\beta\alpha}}\Gamma_{1(2)} \qquad (1.5.29)$$

We examine a case in which $P_i = \mu_i$:

$$P^{\alpha} = \mu_2 n_2^{\alpha} + \mu_1 n_1^{\alpha} = g_v^{\alpha} \qquad (1.5.30)$$

$$P^{\beta} = \mu_2 n_2^{\beta} + \mu_1 n_1^{\beta} = g_v^{\beta} \qquad (1.5.31)$$

where g_v^{α} and g_v^{β} are the densities of the Gibbs thermodynamic potential in the phases α and β:

$$\Gamma_{1(\mu)} = -\frac{\mu_2\left(n_2^{\alpha} - n_1^{\alpha}\right)}{(g_v^{\alpha} - g_v^{\beta})}\left(\frac{\partial\sigma}{\partial\mu_1}\right) \qquad (1.5.32)$$

$$S_{\omega(\mu)} = -\left(\frac{\partial\sigma}{\partial \mathrm{T}}\right)_{\mu_1} + \frac{\mu_2\left(s^{\alpha} - s^{\beta}\right)}{\left(g_v^{\alpha} - g_v^{\beta}\right)}\left(\frac{\partial\sigma}{\partial\mu_1}\right) \qquad (1.5.33)$$

$$S_{\omega(\mu)} = S_{\omega(2)} - \frac{\mu_i\left(s^{\beta} - s^{\alpha}\right)}{g_v^{\alpha} - g_v^{\beta}}\Gamma_{1(2)} \qquad (1.5.34)$$

To use the last two equations for the calculations, it is necessary to have data on the absolute value of the chemical potentials of the component, and experiments only yield the variation of the chemical potential. Therefore, these equations can be used for calculations in practice. At the same time, the expressions (1.5.32)–(1.5.34) are interesting because they determine the relationship of the excess values in relation to the equienergetic dividing surface with the variation of surface tension with composition and temperature.

Thus, the resultant thermodynamic relationships may be used for calculating the integral excess values of the fracture surface, if the dependence of surface tension and chemical potentials on temperature and concentration has been determined by experiments or theoretically.

In conclusion, it is necessary to determine the thermodynamic systems for which these relationships are valid. When deriving these relationships, no approximations or assumptions were made in relation to the structure and composition of the phases. Therefore, these equations are as general as the Gibbs adsorption equation. However, it must be remembered that the adsorption equation was derived by Gibbs for fluid phases. Therefore, these relationships can also be used only for interfaces such as liquid–gas, liquid–liquid, and liquid–plasma.

The resultant relationships are greatly simplified in cases in which it is possible to ignore the content of the components in one of the phases. This assumption is fulfilled quite accurately for a large group of two-phase systems that are important for practice. Therefore, this case will be the subject of a special section.

1.6. Adsorption in two-phase systems with a low content of components in one of the phases

As shown in previous considerations, the adsorption in Gibbs theory of capillarity is the 'excess' quantity characterising the deviation of the real two phase system from the reference system in which the properties of the phases remain constant up to the dividing surface. The value of adsorption depends on where the dividing surface is hypothetically generated. It is therefore necessary to examine the feasibility of using this parameter in thermodynamic calculations. However, the results of these calculations provide objective information on the distribution of components between the coexisting phases and fracture surface. On the other hand, in cases in which it is possible to ignore the content of the components in one of the coexisting phases, the value of adsorption may be directly linked with the real physical quantities, characterising the composition of the surface layer averaged out in respect of the entire region of heterogeneity. The adsorption on different dividing surfaces acquires a specific physical meaning.

This question was examined in detail in [22] and in this book we confine ourselves only to a brief examination of the main moments.

Let the density of the number of the particles in the β-phase to be considerably smaller than that in the α-phase, i.e. $n^\alpha \gg n^\beta$. Therefore,

$$\sum_{i=1}^{n} N_i^\beta \ll \sum_{i=1}^{n} N_i^\alpha \tag{1.6.1}$$

and the equation for adsorption is simplified:

$$\Gamma_i = \frac{N_i - N_i^\alpha - N_i^\beta}{A} = \frac{N_i - N_i^\alpha}{A} \tag{1.6.2}$$

As shown below, the resultant equation provides a physical interpretation of the value of adsorption on different dividing surfaces.

1. Relative Gibbs adsorption in relation to the j-th component ($\Gamma_\gamma = 0$):

$$\Gamma_j = \frac{N_j - N_j^\alpha}{A} = 0 \qquad (1.6.3)$$

Consequently, $N_i = N_j^\alpha$. The last condition show that the quantity Γ_i (γ) maybe represented as is the difference of the number of moles of the i-th component in the fracture surface of the unit area and in some volume of the α-phase which contains the same number of the moles of the j-th component as the examined part of the fracture surface.

2. Adsorption on the equimolar dividing surface (N is a variant of adsorption according to Guggenheim and Adam). The position of the equimolar dividing surface is determined by the condition:

$$\sum_{i=1}^{n} \Gamma_{i(n)} = 0. \qquad (1.6.4)$$

Substituting (1.6.2) into (1.6.4) gives:

$$\sum_{i=1}^{n} \Gamma_{i(n)} = \sum_{i=1}^{n} \frac{N_i - N_i^\alpha}{A} = \frac{\sum_{i=1}^{n} N_i - \sum_{i=1}^{n} N_i^\alpha}{A} = 0 \qquad (1.6.5)$$

$$\sum_{i=1}^{n} N_i = \sum_{i=1}^{n} N_i^\alpha = N \qquad (1.6.6)$$

Taking into account (1.6.6) gives:

$$\Gamma_{i(n)} = \frac{N_i - N_i^\alpha}{A} = \frac{\frac{N_i}{N} - \frac{N_i^\alpha}{N}}{A/N} = \frac{x_i^\omega - x_i^\alpha}{\omega} \qquad (1.6.7)$$

where x_i^ω is the molar fraction of the i-th component, averaged-out in respect of the entire thickness of the fracture surface; x_i^α is the molar fraction of the i-th component in the phase α; ω = A / N is the area of the fracture surface per 1 mol of the substance (molar area).

Condition (1.6.6) indicates that we have compared the volumes with the same total number of the moles of all components. Consequently, the adsorption of the i-th component in relation to the equimolar dividing surface is numerically equal to the difference of the number of moles of the i-th component in the volume of the liquid, adjacent to the unit surface, and in the volume in the depth of the solution containing the same total number of moles of all components.

3. Adsorption on the physical dividing surface. According to the definition, the position of the physical dividing surface is 'fixed' by the condition:

$$\sum_{i=1}^{n}\Gamma_{i(V)}\overline{V}_i^{\alpha}=0 \tag{1.6.8}$$

Substituting (1.6.2) we obtain:

$$\sum_{i=1}^{n}\frac{N_i-N_i^{\alpha}}{A}\overline{V}_i^{\alpha}=\sum_{i=1}^{n}\frac{N_i\overline{V}_i^{\alpha}-N_i^{\alpha}\overline{V}_i^{\alpha}}{A}=0; \tag{1.6.9}$$

$$\sum_{i=1}^{n}N_i\overline{V}_i^{\alpha}=\sum_{i=1}^{n}N_i^{\alpha}\overline{V}_i^{\alpha}$$

or taking into account that $\sum_{i=1}^{n}N_i\overline{V}_i^{\alpha}=V^{\alpha}$ and $\sum_{i=1}^{n}N_i\overline{V}_i^{\alpha}=V^{\omega}$:

$$V^{\omega}=V^{\alpha} \tag{1.6.10}$$

As indicated by (1.6.10), the adsorption of the physical dividing surface is the difference of the number of the moles of the i-th component in the volume, adjacent to the unit surface, and the number of the moles of the component in the same volume in the depth of the solution.

It is interesting to examine the term 'the physical dividing surface'. This term reflects the situation in the liquid–vapour two-phase system, situated away from the critical temperature. In this case, it may be assumed that there are two areas of the variation of the local properties. In one of them, this variation is determined by the substitution of molecules of one type by molecules of another type whilst retaining the volume per molecule in the solution, *i.e.*, without changing the molecular spacing. If this assumption is close to the actual situation, then ignoring the mass of the vapour part of the fracture surface, we obtain the dividing surface whose position is almost identical with the physical phase boundary, *i.e.*, the surface is situated in the section of the rapid variation of the local concentration. Therefore, the adsorption on the dividing surface, given by the condition (1.6.8), is referred to as the adsorption on the physical dividing surface.

The expression for adsorption at the physical phase boundary has the form:

$$\Gamma_{i(V)}\overline{V}_i^{\alpha} = \frac{N_i\overline{V}_i^{\alpha} - N_i^{\alpha}\overline{V}_i^{\alpha}}{A} = \frac{v_i^{\omega} - v_i^{\alpha}}{A}, \tag{1.6.11}$$

where $v_i^{\omega}, v_i^{\alpha}$ are the molar volume of the i-th component in the fracture surface and in the α-phase, respectively.

We divide the numerator and denominator by V^{α}, *i.e.*, the volume of the compared section of the fracture surface:

$$\Gamma_{i(V)}V_i^{\alpha} = \frac{y_i^{\omega} - y_i^{\alpha}}{A/V^{\alpha}} = \frac{V^{\alpha}}{A}\left(y_i^{\omega} - y_i^{\alpha}\right) \tag{1.6.12}$$

Taking into account that $\frac{V^{\alpha}}{A} = \frac{A\tau}{A} = \tau$, we obtain

$$\Gamma_{i(V)} = \tau(y_i^{\omega} - y_i^{\alpha}); \tag{1.6.13}$$

where y_i^{ω} is the averaged-out volume fraction of the i-th component in the fracture surface, y_i^{α} is the volume fraction of this component in the α-phase.

Thus, the adsorption of the physical dividing surface is directly linked with the difference of the volume fractions of the component in the surface and in the α-phase.

4. Adsorption on the equimass dividing surface (M-variant according to Guggenheim and Adam)

The position of the equimass dividing surface is determined by the condition $\sum_{i=1}^{n}\Gamma_{i(M)}M_i = 0$. Substituting instead of adsorption its value (1.6.2), after several transformations we obtain:

$$\sum_{i=1}^{n}\frac{N_i - N_i^{\alpha}}{A}M_i = 0 \tag{1.6.14}$$

or

$$m^{\omega} = m^{\alpha} \tag{1.6.15}$$

Taking (1.6.15) into account

$$\Gamma_{i(M)}M_i = \frac{N_iM_i - N_i^{\alpha}M_i}{A} = \frac{m_i^{\omega} - m_i^{\alpha}}{A} = \frac{c_i^{\omega} - c_i^{\alpha}}{A/m^{\sigma}} \tag{1.6.16}$$

Since $\frac{A}{m^{\omega}} = \frac{A}{M\nu} = \frac{\omega}{M}$, where M is the molar mass, ν is the number of moles, the expression for $\Gamma_{i(M)}$ may be presented in the following form:

$$\Gamma_{i(M)} = M\frac{c_i^{\omega} - c_i^{\alpha}}{\omega} \tag{1.6.17}$$

where ω is the area per 1 mol of the substance in the fracture surface; c_i^ω is the averaged-out mass fraction of the i-th component in the fracture surface; c_i^α is the mass fraction of the i-th component in the α-phase. Thus, the adsorption on the equimass dividing surface is the difference of the number of moles of the i-th component in the mass of the fracture surface of the unit area and in the same mass in the depth of the solution. On the other hand, the expression (1.6.17) links the sorption on the equimass surface with a difference of the mass fraction of the component in the fracture surface and in the α-phase. These considerations show that the molar area is present in almost all expressions.

We examine the procedure of evaluating the area per 1 mol of substance in the interfacial layer.

Let the thickness of the interfacial layer be equal to the thickness of the monolayer of the particles (atoms or molecules). Consequently, the mean distance between the particles is:

$$l = \left(v^\omega\right)^{\frac{1}{3}} \tag{1.6.18}$$

where $v^\omega = \dfrac{M}{\rho^\omega N_A}$ is the volume per particle in the interfacial layer, M is the mass 1 mol of substance, ρ^ω is the mean density of the interfacial layer, N_A is the Avogadro number.

Consequently in the interfacial layer, one particle is situated on average on the area:

$$l^2 = \left[\frac{M}{N_A \rho^\omega}\right]^{\frac{2}{3}} \tag{1.6.19}$$

and the mole of substance occupies the area:

$$\omega = N_A l^2 = N_A \left(\frac{M}{N_A \rho^\omega}\right)^{\frac{2}{3}} = N_A^{1/3}\left(\frac{M}{\rho^\omega}\right)^{2/3} \tag{1.6.20}$$

When the thickness of the interfacial layer is greater than the thickness of the monolayer (interlayer) and amounts to n-monolayers, the last equation should be divided by the number of monolayers:

$$\omega = \frac{1}{n} N_A^{1/3}\left(\frac{M}{\rho^\omega}\right)^{2/3} \tag{1.6.21}$$

The mean density of the surface layer ρ^ω is an undetermined quantity and, therefore, the so-called packing coefficient $f=\left(\rho^{\alpha}/\rho^{\omega}\right)^{\frac{2}{3}}$ is introduced, where ρ^{α} is the density of the α-phase.

Taking into account the introduced notation, the expression for the molar area will be converted to the form:

$$\omega=\frac{f}{n}N_A^{1/3}\left(\frac{M}{\rho}\right)^{\frac{2}{3}}$$

In this equation, ρ is the density of one of the volume phases and this simplifies calculation. The values of the parameters f and n are usually estimated on the basis of the nature of the coexisting phases. In particular, for the liquid–vapour interface away from the critical temperature, the values of f and n are close to 1. As an example, it should be mentioned that in the case in which ρ^{ω} = 0.95 ρ_l, f ≈ 1.03.

Thus, the resultant equations show that only the adsorption on the molar dividing surface is directly linked with the difference of the molar fractions of the component in the fracture surface and in the α-phase. However, to calculate the averaged-out molar fraction of the component in the fracture surface, it is necessary to know the thickness of the surface or the number of monolayers which the fracture surface includes.

1.7. Calculation of adsorption at the liquid–vacuum interface on the basis of the concentration dependence of surface tension

In section 1.5 equations were derived for calculating the adsorption on different dividing surfaces of any two-component, two-phase systems. These equations are greatly simplified in cases in which it is possible to ignore the content of the component in one of the phases, i.e. the condition $\rho^{\beta}\ll\rho^{\alpha}$ is fulfilled. Actually, in this case, we have $n_i^{\beta}\ll n_i^{\alpha}$, and the equations (1.5.12)–(1.5.34) can be transformed and presented in the following form:

$$\Gamma_{1(2)}=-\frac{a_1}{RT}\left(\frac{\partial\sigma}{\partial a_1}\right)_T;\quad S_{\omega(2)}=-\left(\frac{\partial\sigma}{\partial \mathrm{T}}\right)_{\mu_1};\qquad(1.7.1)$$

$$\Gamma_{i(N)}=-\frac{\left(n_2^{\alpha}-n_2^{\beta}\right)a_1}{\left(n^{\alpha}-n^{\beta}\right)RT}\left(\frac{\partial\sigma}{\partial a_1}\right)=-\frac{n_2^{\alpha}a_1}{n^{\alpha}RT}\left(\frac{\partial\sigma}{\partial a_1}\right)_{\mathrm{T}}\qquad(1.7.2)$$

Taking into account that $\frac{n_2^\alpha}{n^\alpha} = x_2^\alpha = \left(1 - x_1^\alpha\right)$,

$$\Gamma_{i(n)} = -\frac{\left(1 - x_1\right) a_1}{RT} \left(\frac{\partial \sigma}{\partial a_1}\right) \tag{1.7.3}$$

$$S_{\omega(n)} = -\left(\frac{\partial \sigma}{\partial \mathrm{T}}\right)_{\mu_2} + \frac{a_1 \left(s^\alpha - s^\beta\right)}{RTn^\alpha} \left(\frac{\partial \sigma}{\partial a_1}\right)_{\mathrm{T}} \tag{1.7.4}$$

or

$$S_{\omega(n)} = -S_{\omega(2)} - \left(s^\alpha - s^\beta\right) \frac{\Gamma_{1(2)}}{n^\alpha} \tag{1.7.5}$$

Similarly, for other methods of determination of adsorption

$$\Gamma_{i(M)} = -\frac{M_2 \left(n_2^\alpha - n_2^\beta\right) a_1}{\left(\rho^\alpha - \rho^\beta\right) RT} \left(\frac{\partial \sigma}{\partial a_1}\right)_{\mathrm{T}} = -\frac{M_2 n_2^\alpha a_1}{\rho^\alpha RT} \left(\frac{\partial \sigma}{\partial a_1}\right) \tag{1.7.6}$$

Since $\rho^\alpha = M_2 n_2^\alpha + M_1 n_1^\alpha$, then

$$\Gamma_{i(M)} = -\frac{a_1}{\left(1 + \frac{M_1 n_1^\alpha}{M_2 n_2^\alpha}\right) RT} \left(\frac{\partial \sigma}{\partial a_1}\right)_T \tag{1.7.7}$$

$$\Gamma_{i(M)} = -\frac{a_1}{\left(1 + \frac{M_1 x_1^\alpha}{M_2 x_2^\alpha}\right) RT} \left(\frac{\partial \sigma}{\partial a_1}\right)_T \tag{1.7.8}$$

$$S_{\omega(M)} = S_{\omega(2)} - \left(s^\alpha - s^\beta\right) \frac{M_1 \Gamma_{1(2)}}{\rho^\alpha} \tag{1.7.9}$$

$$\Gamma_{1(V)} = -\frac{\overline{V}_2^\alpha \left(n_2^\alpha - n_2^\beta\right) a_1}{(\overline{V}^\alpha - \overline{V}^\beta) RT} \left(\frac{\partial \sigma}{\partial a_1}\right) = \frac{-a}{\left(1 - \overline{V}_1^\alpha n_1^\alpha / \overline{V}_2^\alpha n_1^\alpha\right) RT} \left(\frac{\partial \sigma}{\partial a_1}\right)_T \tag{1.7.10}$$

$$S_{\omega(V)} = S_{\omega(2)} - \left(s^\alpha - s^\beta\right) \frac{\Gamma_{1(2)}}{1 + \frac{\overline{V}_1^\alpha n_1^\alpha}{\overline{V}_2^\alpha n_2^\alpha}} = \left(s^\alpha - s^\beta\right) \frac{\overline{V}_2^\alpha n_2^\alpha \Gamma_{1(2)}}{V^\alpha} \tag{1.7.11}$$

$$S_{\omega(\mu)} = S_{\omega(2)} + \frac{\mu_2}{g_v^\alpha - g_v^\beta}\left(s^\beta - s^\alpha\right)\Gamma_{1(2)} \tag{1.7.12}$$

$$\Gamma_{1(\mu)} = -\frac{\mu_2\left(n_2^\alpha - n_1^\alpha\right)}{g_v^\alpha - g_v^\beta}\left(\frac{\partial \sigma}{\partial \mu_2}\right)_T \tag{1.7.13}$$

Thus, if the concentration and temperature dependences of the surface tension and thermodynamic activities of the two-component systems are determined by experiments, these equations may be used to calculate the adsorption and surface entropy on different dividing surfaces. These data may also be used to determine the distance between different dividing surfaces.

The resultant equations are simple for ideal solutions. Let us assume that we have 'the ideal solution–natural vapour' two-phase, two-component system. The chemical potential of the component in this system is determined by the expression $\mu_i = \mu_{i(0)} + RT\ln x_i$. Consequently at $T =$ const, $d\mu_i = RTd\ \ln x_i$, and the expressions for adsorption may be presented in the following form:

$$\Gamma_{1(2)} = -\frac{x_1}{RT}\left(\frac{\partial \sigma}{\partial x_1}\right)_T \tag{1.7.14}$$

$$\Gamma_{1(n)} = -\frac{x_1\left(1 - x_1\right)}{RT}\left(\frac{\partial \sigma}{\partial x_1}\right)_T \tag{1.7.15}$$

$$\Gamma_{1(V)} = -\frac{x_1}{1 + \dfrac{V_1^\alpha n_1^\alpha}{V_2^\alpha n_2^\alpha}}\left(\frac{\partial \sigma}{\partial x_1}\right)_T \tag{1.7.16}$$

$$\Gamma_{1(\mu)} = -\frac{\mu_2\left(n_2^\alpha - n_1^\alpha\right)}{g_v^\alpha - g_v^\beta}\left(\frac{\partial \sigma}{\partial x_1}\right)_T \tag{1.7.17}$$

The ideal solutions here and in further considerations are the solutions in which the activity coefficient of the components $f_1 = 1$ in the entire composition range.

Similarly, the equations for excess entropy may also be written in a simpler form:

$$S_{\omega(2)} = -\left(\frac{\partial \sigma}{\partial \mathrm{T}}\right)_{X_1}; \tag{1.7.18}$$

$$S_{\omega(n)} = -\left(\frac{\partial \sigma}{\partial \mathrm{T}}\right)_{X_1} + \frac{x_1}{RT}\left(\frac{\partial \sigma}{\partial x_1}\right)_T \frac{s^\alpha - s^\beta}{n^\alpha} \quad (1.7.19)$$

$$S_{\omega(M)} = -\left(\frac{\partial \sigma}{\partial \mathrm{T}}\right)_{X_1} + \frac{x_1}{RT}\left(\frac{\partial \sigma}{\partial x_1}\right)_T \frac{M_1(s^\alpha - s^\beta)}{M_1 n_1^\alpha + M_2 n_2^\alpha} \quad (1.7.20)$$

$$S_{\omega(V)} = -\left(\frac{\partial \sigma}{\partial \mathrm{T}}\right)_{X_1} + \frac{x_1}{RT}\left(\frac{\partial \sigma}{\partial x_1}\right)_T \frac{\overline{V}_1^\alpha (s^\alpha - s^\beta)}{n_1^\alpha \overline{V}_1^\alpha + n_2^\alpha \overline{V}_2^\alpha} \quad (1.7 .21)$$

As indicated by the equations, the sign of adsorption on any dividing surface of a real two component system is determined by the partial derivative $\left(\frac{\partial \sigma}{\partial \mu_i}\right)_T$, and in the case of the ideal systems by the derivative $\left(\frac{\partial \sigma}{\partial x_1}\right)_T$. This derivative characterises the dependence of surface tension on concentration and Rebinder, taking into account the significance of the estimation of the sign of adsorption, refers to it as the surface activity of the component in the solution. The limiting surface activity is the limit

$$G_{0i} = \lim_{x_1 \to 0}\left(\frac{\partial \sigma}{\partial x_1}\right)_T \quad (1.7.22)$$

which characterises the effect of the small additions of the i-th component (admixture) on the surface tension of the main component (solvent).

It should be mentioned that even in real solutions, diluted in respect of the i-th component, expression (1.7.22) remains valid because the condition $\gamma_1 = 1$ is fulfilled in the diluted solutions and

$$\lim_{X_1 \to 0}\left(\frac{\partial \sigma}{\partial a_1}\right)_T = \lim_{X_1 \to 0}\left(\frac{\partial \sigma}{\partial x_i}\right)_T$$

Thus, the partial derivative $\left(\frac{\partial \sigma}{\partial x_i}\right)_T$ at low concentrations x_i determine the sign of adsorption, *i.e.*, makes it possible to estimate the excess or shortage of the given component in the surface layer in comparison with its content in the volume of the solution. If $\left(\frac{\partial \sigma}{\partial x_i}\right)_T \prec 0$ then independently of the method of determination of $\Gamma_i \succ 0$ (the excess of substance on the fracture surface), and this component is referred to as surface active. If $\left(\frac{\partial \sigma}{\partial x_i}\right)_T \succ 0$, then $\Gamma_i \succ 0$, there is a shortage of the component on the fracture surface and the component is referred to as inactive.

Surface-active substances (SAS) are used in solving many technical problems and it is therefore not surprising that special attention has been paid to examining their properties.

To solve a number of problems in practice, it will be necessary to carry out the quantitative estimation of the surface (sometimes referred to as capillary) activity of a component in a solution. Therefore, various criteria and estimates of the surface activity of elements on the basis of other physical–chemical properties have been proposed. Without discussing these criteria in detail, it should be mentioned that none of them predicts the sign of the derivative for all binary systems. These concepts when used for multicomponent systems will be discussed again at the end of this chapter.

Thus, in binary systems, regardless of the determination method, the son of adsorption is determined by a single partial derivative, which characterises the surface activity of the component.

In scientific literature, special attention is often given the problem of advantages of a specific method of determination of adsorption. In some studies [298] there are erroneous claims on the incorrectness of calculating the Gibson relative adsorption throughout the entire range of the composition because the conditions $\Gamma_j = 0$ is fulfilled only in diluted solutions. It has been shown that the dividing surface can always be selected to ensure that the adsorption of a component is equal to zero.

According to some of the authors, the dividing surface whose position does not change with the variation of composition is regarded as suitable.

In the liquid–vapour system, the physical dividing surface, given by the condition $\sum_{i=1}^{n}\Gamma_{i(v)}\overline{V}_i^{\alpha}=0$ is close to the physical boundary and, consequently, its position is also constant with the variation of composition.

On the other hand, in the determination of the mean composition of the fracture surface, expressed in molar fractions, it is more convenient to examine the adsorption on the equimolar dividing surface whose position is given by the condition $\sum_{i=1}^{n}\Gamma_{i(n)}=0.$

These considerations show that in the thermodynamic respect, all the methods of determination of adsorption are equally important and differ from each other only by the method of 'fixing' the dividing surface. Therefore, the selection of the position of the dividing surface should be determined by the efficiency of solving the given task. As mentioned previously, when solving certain theoretical problems, the dividing surface is also selected in relation to other excess quantities, in particular, it is selected to ensure that the excess internal energy is ($U^{\omega} = 0$) or excess entropy is equal to zero ($S^{\omega} = 0$) [26]. It is interesting to also consider the previously examined equienergetic dividing surface $\sum_{i=1}^{n}\Gamma_i\mu_i = 0$ because the condition $f_{\omega} = \sigma$ is fulfilled only in relation to this surface, *i.e.*, the specific free surface energy is equal to surface tension.

This paragraph was concerned with the two phase two-component systems. The solution of these problems for a multicomponent system is associated with certain difficulties examined in the following section.

1.8. Calculation of adsorption in multicomponent systems

When solving many technological problems, it is necessary to solve problems associated with the effect of admixtures on the surface properties of the materials. The admixtures of surface-active substances may greatly change, even in small amounts, the entire course of the adsorption phenomena in binary systems. Therefore, it is very important to examine the problem of the simultaneous effect of several admixtures on the surface tension and adsorption of binary solutions, *i.e.*, examine the surface properties of multicomponent systems.

Many phenomena and processes, in particular those associated with the effect of admixtures on the physical–chemical properties of the interphase layer, may be understood and, consequently, controlled if the relationships governing adsorption phenomena in multicomponent systems are examined. Evidently, this may also be used to explain the special interest showed in recent years by investigators to the examination of the surface properties of multicomponent systems.

In fluid systems, the reliably measure property of the surface is surface tension. Examination of the temperature and concentration dependence of surface tension makes it possible to calculate a number of excess thermodynamic properties of the surface. In this case, as shown previously, the calculation of these quantities in binary systems does not cause any problems, because here it is sufficient to supplement the Gibbs adsorption equation by the condition of fixation of the position of the dividing surface.

An increase of the number of components is accompanied also by an increase of the number of unknown quantities in the Gibbs equation. This greatly complicates the problem of examination and calculation of adsorption on the basis of the concentration dependence of surface tension and chemical potentials. This may also be used to explain the fact that in scientific literature it was believed for a very long period of time that the calculation of adsorption using the Gibbs equation for multicomponent systems is very difficult or even impossible. Various equations have been proposed for approximate calculation of adsorption in multicomponent systems [13, 38]. However, these equations are given for specific models and cannot replace the strict Gibbs thermodynamic equation. Therefore, it is still important to solve this problem for multicomponent systems. The principal possibility of calculating the relative adsorption in multicomponent systems was also shown by Gibbs himself. In further stages, in examining the Gibbs thermodynamic theory, many authors restricted their considerations to the application of the adsorption equation for binary systems. A case of three-component systems was investigated in [6, 19, 20].

This problem will be examined in a general form without making any assumptions regarding the density of the contacting phases. This approach also simplifies examination of simpler cases.

From the Gibbs adsorption equation:

$$-d\sigma=\left(S^{\omega}-\Gamma_j\frac{s^{\alpha}-s^{\beta}}{n_j^{\alpha}-n_j^{\beta}}\right)dT+\sum_{i=1}^{n}\left[\Gamma_i-\Gamma_j\frac{n_i^{\alpha}-n_i^{\beta}}{n_j^{\alpha}-n_j^{\beta}}\right]d\mu_i \qquad (1.8.1)$$

for the dividing surface $\Gamma_j = 0$:

$$-d\sigma=S_{\omega(j)}dT+\sum_{i=1}^{n}\Gamma_{i(j)}\left(1-\delta_{ij}\right)d\mu_i \quad (1\le j\le n) \qquad (1.8.2)$$

where δ_{ij} is the Kronecker symbol, $\Gamma_{i(j)}$ is the adsorption of the i-th component on the dividing surface, corresponding to the zero value of adsorption of the j-th component. As noted by Gibbs, in equation (1.8.2) all the differentials are independent, and if the functional dependence of surface tension on $\mu_1,\dots,\ \mu_{j-1},\ \mu_{j+1},\ \mu_n$, T is available, the following equation may be written for adsorption:

$$\Gamma_{i(j)}=-\left(\frac{\partial\sigma}{\partial\mu_i}\right)_{T,\mu_{k\neq i\neq j}} \qquad (1.8.3)$$

In this equation, the chemical potentials of the i-th and j-th component may change so that it may be useful for calculations in practice. In the case of ideal solutions, taking into account $\mu_i = \mu_{i0} + RTlnx_i$, equation (1.8.3) is converted as follows:

$$\Gamma_{m(j)} = -\frac{x_m}{RT}\left(\frac{\partial \sigma}{\partial x_m}\right)_{x_{k \neq m \neq j}} \tag{1.8.4}$$

An identical equation may also be written for excess entropy:

$$S_{\omega(j)} = -\left(\frac{\partial \sigma}{\partial T_i}\right)_{\mu_{i \neq j}} \tag{1.8.5}$$

It should be noted that these last two equations examine the dependence of the surface tension of the solution on temperature at constant chemical potentials (atomic fractions) of all components. Similarly, we can obtain the expression for excess entropy and adsorption in relation to other dividing surfaces. In fact, taking into account (1.4.5) from (1.8.1):

$$-d\sigma = \left(S_{\omega(P)} + \sum_{\substack{k=1 \\ k \neq j}}^{n} \frac{P_k}{P_j} \frac{\Gamma_{k(P)}\left(s^\alpha - s^\beta\right)}{n_j^\alpha - n_j^\beta}\right) dT + \sum_{\substack{i=1 \\ i \neq j}}^{n} \left(\Gamma_{i(P)} + \sum_{k \neq j}^{n} \frac{P_k}{P_j} \Gamma_{k(P)} \frac{n_i^\alpha - n_i^\beta}{n_j^\alpha - n_j^\beta}\right) d\mu_i \tag{1.8.6}$$

Consequently

$$S_{\omega(P)} = -\left(\frac{\partial \sigma}{\partial T}\right)_{\mu_{i \neq j}} - \sum_{k \neq j}^{n} \frac{P_k \Gamma_{k(P)}}{P_j} \frac{s^\alpha - s^\beta}{n_j^\alpha - n_j^\beta}; \tag{1.8.7}$$

$$\Gamma_{i(P)} + \sum_{k \neq j}^{n} \frac{P_k \Gamma_{k(P)}}{P_j} \frac{n_i^\alpha - n_i^\beta}{n_j^\alpha - n_j^\beta} = -\left(\frac{\partial \sigma}{\partial \mu_i}\right)_{T, \mu_{k \neq j \neq i}} \tag{1.8.8}$$

where $i = 1, 2, \ldots, j-1, j+1, \ldots, n$.

In the last equation, the index i has n-1 values and we have a system of n-1 linear equations

$$\begin{cases} \Gamma_{1(P)} + \sum_{k\neq j}^{n} \frac{P_k \Gamma_{k(P)}}{P_j} \frac{n_1^\alpha - n_1^\beta}{n_j^\alpha - n_j^\beta} = -\left(\frac{\partial \sigma}{\partial \mu_1}\right)_{T,\mu_{k\neq j\neq 1}} \\ \Gamma_{2(P)} + \sum_{k\neq j}^{n} \frac{P_k \Gamma_{k(P)}}{P_j} \frac{n_2^\alpha - n_2^\beta}{n_j^\alpha - n_j^\beta} = -\left(\frac{\partial \sigma}{\partial \mu_2}\right)_{T,\mu_{k\neq j\neq 2}} \\ \cdots\cdots\cdots\cdots\cdots\cdots \\ \cdots\cdots\cdots\cdots\cdots\cdots \\ \Gamma_{n(P)} + \sum_{k\neq j}^{n} \frac{P_k \Gamma_{k(P)}}{P_j} \frac{n_n^\alpha - n_n^\beta}{n_j^\alpha - n_j^\beta} = -\left(\frac{\partial \sigma}{\partial \mu_n}\right)_{T,\mu_{k\neq j\neq n}} \end{cases} \quad (1.8.9)$$

If the derivatives $\left(\frac{\partial \sigma}{\partial \mu_i}\right)_{T,\mu_{k\neq j\neq 1}}$ are determined by experimental theoretically, from the system of equations (1.8.9) we can determine the adsorption of all components. For ideal solutions, these equations are written in the following form:

$$\begin{cases} \Gamma_{1(P)} + \sum_{k\neq j}^{n} \frac{P_k \Gamma_{k(P)}}{P_j} \frac{n_1^\alpha - n_1^\beta}{n_j^\alpha - n_j^\beta} = -\frac{x_1}{RT}\left(\frac{\partial \sigma}{\partial x_1}\right)_{T,x_{k\neq j\neq 1}} \\ \Gamma_{2(P)} + \sum_{k\neq j}^{n} \frac{P_k \Gamma_{k(P)}}{P_j} \frac{n_2^\alpha - n_2^\beta}{n_j^\alpha - n_j^\beta} = -\frac{x_2}{RT}\left(\frac{\partial \sigma}{\partial x_2}\right)_{T,x_{k\neq j\neq 2}} \\ \cdots\cdots\cdots\cdots\cdots\cdots \\ \cdots\cdots\cdots\cdots\cdots\cdots \\ \Gamma_{n(P)} + \sum_{k\neq j}^{n} \frac{P_k \Gamma_{k(P)}}{P_j} \frac{n_n^\alpha - n_n^\beta}{n_j^\alpha - n_j^\beta} = -\frac{x_n}{RT}\left(\frac{\partial \sigma}{\partial x_n}\right)_{T,x_{k\neq j\neq n}} \end{cases} \quad (1.8.10)$$

The equations (1.8.3) and (1.8.9) may in principle be used to calculate the adsorption of components in real multicomponent systems, but they have one common shortcoming. The independent variables in these equations are represented by chemical potentials. In practice, it is difficult to vary the composition of the melt along the lines $\mu_{k\neq i\neq j}$ = const, and this creates difficulties in calculations. Thus, as with binary systems, we transfer from chemical potentials to molar fractions of the components in one of the phases.

However, in contrast to the binary system, the composition of the n-component system characterises $n-1$ variables (molar fractions) which may be changed irrespective of each other. The multicomponent systems contain an infinite number of the methods of variation of composition, whereas only one method may be used in the binary systems. Therefore, when discussing the problem of calculation of adsorption in multicomponent systems, it is necessary to use a different method in comparison with the binary system. In particular, when solving these problems it is necessary to indicate the method of variation of the composition of the multicomponent solution. This is the qualitative difference between the multi-component systems and binary solutions.

In principle, the composition of a multicomponent solution may be varied in any direction. Therefore, specifying the method of variation of the composition of a multicomponent solution in the form of the function F_k = const and, at the same time, imposing additional conditions on the differential equations (1.8.2) and (1.8.4), at T = const we obtain systems of linear equations in which the independent variables are represented by molar fractions:

$$-\left(\frac{\partial \sigma}{\partial x_m}\right)_{F_m} = \sum_{i=1}^{n} \Gamma_{i(j)} \left(1-\delta_{ij}\right)\left(\frac{\partial \mu_i}{\partial x_m}\right)_{F_m} \tag{1.8.11}$$

$m = 1, 2, \ldots, j\text{-}1, j+1, n$

$$-\left(\frac{\partial \sigma}{\partial x_m}\right)_{F_m} = \sum_{i=1}^{n}\left[\Gamma_{i(P)} + \sum_{k\neq j}^{n} \frac{P_k}{P_j} \Gamma_{k(P)} \frac{n_i^{\alpha}-n_i^{\beta}}{n_j^{\alpha}-n_j^{\beta}}\right]\left(\frac{\partial \mu_i}{\partial x_m}\right)_{F_m} \tag{1.8.12}$$

$m = 1, 2, \ldots, j-1, j+1, n$

Index F_m = const indicates the selected method of variation of the composition of the solution.

The systems of the linear equations (1.8.11) and (1.8.12) contain $n-1$ unknown values of adsorption which may be determined using the Kramer rule:

$$\Gamma_{m(j)} = \frac{\Delta m}{\Delta} \tag{1.8.13}$$

$$\Gamma_{m(P)} = \frac{\Delta m}{\Delta} \tag{1.8. 14}$$

where Δ and Δm are the main and auxiliary determinants of the systems of linear equations.

The values of the partial derivatives, included in (1.8.11) and (1.8.12) may be determined if the dependence of the interfacial tension and chemical potentials on the composition of one of the phases is known. Consequently, solving these systems of the linear equations, it is also easy to determine the adsorption values of the component. It should be mentioned that when deriving (1.8.11) and (1.8.12) no assumptions were made regarding the structure and properties of the phases. Therefore, they are as general as the Gibbs adsorption equation.

To calculate adsorption using equations (1.8.11) and (1.8.12), it is necessary to select a methods of changing the composition and have data on the dependence of surface tension and chemical potential on the composition of one of the phases. The equations are general and hold for any method of variation of the composition of the multicomponent solution. However, two methods of changing the composition of the multicomponent system are more suitable for practical calculations:

1. The composition of a multicomponent solution may be changed in such a manner that the molar fractions of all components, with the exception of two components, remain constant, *i.e.*, the atom of one component in the solution replaced by the atoms of another component:

$$F_k = x_k = cons\ t,\quad k \neq j \neq m \tag{1.8.15}$$

In this case, only the molar fraction of the j-th and m-th component may change. This method of changing the composition may be used to examine the variation of the type of isotherms of the physical–chemical properties of the binary system with the increase of the content of the third component.

2. When the composition of the solution varies, the ratio of the molar fractions of all components with the exception of one should remain constant:

$$F_k = \frac{x_i}{x_r} = \text{const},\ r,\ i \neq m \tag{1.8.16}$$

This method of variation of the composition corresponds to the addition of a component denoted by index m. In the case of ternary systems, this is a family of secants, extending from the tips of the Gibbs concentration triangle. These variants will be examined separately.

1.9. Calculation of the adsorption of components with the variation of the composition along lines with a constant content of components, with the exception of two components

It is assumed that the composition of the solution changes in such a manner that the molar fraction of all components, with the exception of two, remain unchanged, *i.e.*, in the solution, the atoms of the i-th component are replaced by the atoms of the j-th component: $F_k = x_{k \neq i \neq j}$ = const. In this case, only the molar fractions of the j-th and i-th components may change.

In the case of a ternary system, it is the variation of composition at constant molar fractions of one component, *i.e.*, along the lines parallel to the sides of the Gibbs concentration triangle. Consequently, the equations (1.8.11) and (1.8.112) are presented in a new form:

$$-\left(\frac{\partial \sigma}{\partial x_m}\right)_{x_{k \neq m \neq j}} = \sum_{i=1}^{n} \Gamma_{i(j)} \left(1 - \delta_{ij}\right) \left(\frac{\partial \mu_i}{\partial x_m}\right)_{x_{k \neq m \neq j}} \tag{1.9.1}$$

$$-\left(\frac{\partial \sigma}{\partial x_m}\right)_{x_{k \neq m \neq j}} = \sum_{i=1}^{n} \left[\Gamma_{i(P)} + \sum_{k=j}^{n} \frac{P_k}{P_j} \Gamma_{k(P)} \frac{n_i^{\alpha} - n_i^{\beta}}{n_j^{\alpha} - n_j^{\beta}}\right] \left(\frac{\partial \mu_i}{\partial x_m}\right)_{x_{k \neq m \neq j}} \tag{1.9.2}$$

$m = 1, 2, \ldots, j-1, j+1, n$

In the resultant equations, m has $n-1$ values and the expressions (1.9.1) and (1.9.2) represent a system of linear equations in relation to the adsorption of components. In the presence of the dependence of surface tension and chemical potentials on composition, these equations may be used to determine the adsorption of components in relation to the known positions of the dividing surface.

However, although they may be used for calculating adsorption on any dividing surface of a real of two-phase system, the resultant systems of the equations in the absence of data on the functional dependence of the chemical potentials on composition are not suitable for theoretical analysis of adsorption phenomena in real multicomponent systems. Therefore, attention will be given to the case of ideal solutions ($\mu_i = \mu_{i0} + RT\ln x_i$, $d\mu_i = RTd\ln x_i$) for which equations (1.9.1) and (1.9.2) give:

$$-\frac{1}{RT}\left(\frac{\partial\sigma}{\partial x_m}\right)_{x_{k\neq m\neq j}}=\sum_{i=1}^{n}\Gamma_{i(j)}\left(1-\delta_{ij}\right)\left(\frac{\partial\ln x_i}{\partial x_m}\right)_{x_{k\neq m\neq j}} \tag{1.9.3}$$

$m = 1, 2, \ldots, j-1, j+1, n;$

$$-\frac{1}{RT}\left(\frac{\partial\sigma}{\partial x_m}\right)_{x_{k\neq m\neq j}}=\sum_{i=1}^{n}\left[\Gamma_{i(P)}+\sum_{k=j}^{n}\frac{P_k}{P_j}\Gamma_{k(P)}\frac{n_i^\alpha-n_i^\beta}{n_j^\alpha-n_j^\beta}\right]\left(\frac{\partial\ln x_i}{\partial x_m}\right)_{x_{k\neq m\neq j}} \tag{1.9.4}$$

$m = 1, 2, \ldots, j-1, j+1, n.$

It is evident that at $i\neq m\neq j$:

$$\left(\frac{\partial\ln x_i}{\partial x_m}\right)_{x_{k\neq m\neq j}}=\begin{cases}0 & \text{for } i\neq m\neq j;\\ \dfrac{1}{x_m} & \text{for } i=m.\end{cases} \tag{1.9.5}$$

Substituting (1.9.5) into (1.9.3) and (1.9.4) and solving the system of linear equations give:

$$\Gamma_{i:(j)}=-\frac{x_m}{RT}\left(\frac{\partial\sigma}{\partial x_m}\right)_{x_{k\neq m\neq j}} \tag{1.9.6}$$

$$\Gamma_{m(P)}=\frac{Y_m}{RT\sum_{i=1}^{n}P_kY_k}\sum_{i=1}^{n}P_ix_i\left(\frac{\partial\sigma}{\partial x_i}\right)_{x_{k\neq i\neq j}}\left(1-\delta_{ij}\right)-\frac{x_m}{RT}\left(\frac{\partial\sigma}{\partial x_m}\right)_{x_{m\neq i\neq j}} \tag{1.9.7}$$

where $Y_i=n_i^\alpha-n_i^\beta$.

Equation (1.9.7) greatly is simplified if the content of components in one of the volume phases is ignored:

$$\Gamma_{m(P)}=-\frac{x_m}{RT}\left(\frac{\partial\sigma}{\partial x_k}\right)_{x_{k\neq m\neq j}}+\frac{x_m}{P}\sum_{\substack{k=1\\k\neq j}}^{n}P_k\frac{x_k}{RT}\left(\frac{\partial\sigma}{\partial x_k}\right)_{x_{i\neq k\neq j}} \tag{1.9.8}$$

where $P=\sum_{k=1}^{n}P_kx_k$.

As indicated by the equations (1.9.7) and (1.9.8), in ideal solutions adsorption is associated by a simple equation with the derivatives $\left(\frac{\partial\sigma}{\partial x_m}\right)_{x_{k\neq m\neq j}}$. However, equation (1.9.7) is simplest in this respect and indicates that the value and sign of relative adsorption, as in binary systems, are determined by a single derivative. In this case, it should be mentioned that in the case of a multicomponent system this derivative is considered for the variation of the composition along the lines with a constant content of all components, except two.

In conclusion, it should be mentioned that the expression for the relative Gibbs adsorption (1.9.7) does not include the value of the density of the number of particles and the coexisting phases, *i.e.*, this equation is suitable for liquid–liquid and liquid–vapour two-phase systems. Possibly, this may be another advantage of the Gibbs variant and this fact has a deeper physical meaning than simple feasibility for practical application.

Thus, the examined methods of variation of the composition should be examined in calculations of the relative Gibbs adsorption. This gives simple equations (1.9.7) linking the adsorption with a single partial derivative $\left(\frac{\partial\sigma}{\partial x_m}\right)_{x_{k\neq j\neq i}}$. However, this methof of the variation of the composition leads to complicated equations for the Guggenheim and Adam variants. As indicated by (1.9.8), to calculate the adsorption of the component, it is necessary to find $n-1$ values of partial derivatives.

As shown later, these equations are greatly simplified in examining the second method of variation of the composition.

1.10. Calculation of adsorption of components for the variation of the composition by adding one of the components

The second method of variation of the composition will be examined. Let the composition of the solution to change along being sections so that $K_{jr}=\frac{x_j}{x_r}=\text{const}, j,\ r\neq m$. Therefore, the equations (1.8.11) and (1.8.12) are presented in the following new form:

$$-\left(\frac{\partial\sigma}{\partial x}\right)_{K_{ir}}=\sum_{i=1}^{n}\Gamma_{i(j)}\left(1-\delta_i\right)\left(\frac{\partial\mu_i}{\partial x_m}\right)_{K_{i,r}} \qquad (1.10.1)$$

$m = 1, 2, \ldots, j-1, j+1, n.$

$$-\left(\frac{\partial\sigma}{\partial x_m}\right)_{K_{i,r}}=\sum_{i\neq j}^{n}\left[\Gamma_{i(P)}+\sum_{i\neq j}^{n}\frac{P_k}{P_j}\Gamma_{k(P)}\frac{n_i^\alpha-n_i^\beta}{n_j^\alpha-n_j^\beta}\right]\left(\frac{\partial\mu_i}{\partial x_m}\right)_{K_{i,r}} \qquad (1.10.2)$$

$m = 1, 2, ..., j-1, j+1, n$.
In these equations, m has $n-1$ values and the expressions (1.10.1) and (1.10.2) represent a system of linear equations in relation to the adsorption of the components. In the presence of the dependence of surface tension and chemical potentials on composition, these systems of equations may be used to determine the adsorption of components in real multicomponent systems in relation to the known positions of the dividing surface.

In ideal solutions, using the expressions for the chemical potentials, it is possible to determine the values of the partial derivatives of the chemical potential for molar fractions $\left(\frac{\partial \mu_i}{\partial x_m}\right)$. Consequently, the system of the linear equations can be solved giving explicit expressions for adsorption. In fact, transferring to ideal solutions, equations (1.10.1) and (1.10.2) give:

$$-\frac{1}{RT}\left(\frac{\partial \sigma}{\partial x_m}\right)_{K_{i,r}} = \sum_{i=1}^{n} \Gamma_{i(j)}\left(1-\delta_{ij}\right)\left(\frac{\partial \ln x_i}{\partial x_m}\right)_{K_{i,r}} \tag{1.10.3}$$

$$-\frac{1}{RT}\left(\frac{\partial \sigma}{\partial x_m}\right)_{K_{i,r}} = \sum_{\substack{i=1\\ i\neq j}}^{n}\left[\Gamma_{i(P)} + \sum_{k\neq j}^{n} \frac{P_k}{P_j}\Gamma_{k(P)} \frac{n_i^{\alpha}-n_i^{\beta}}{n_j^{\alpha}-n_j^{\beta}}\right]\left(\frac{\partial \ln x_i}{\partial x_m}\right)_{K_{i,r}} \tag{1.10.4}$$

It may easily be shown that the partial derivatives in the equations (1.10.3) and (1.10.4) are determined by the expressions:

$$\left(\frac{\partial \ln x_i}{\partial x_m}\right)_{k_{ir}} = \begin{cases} \dfrac{1}{x_m - 1} & \text{for } i \neq m; \\[2ex] \dfrac{1}{x_m} & \text{for } i = m. \end{cases} \tag{1.10.5}$$

In fact, taking into account that the composition changes at $K_{ir} = \frac{x_i}{x_r} = \text{const}$, from the identity $\sum_{i=1}^{n} x_i = 1$:

$$\sum_{\substack{r=1\\ r\neq m}}^{n} \frac{x_i}{k_{ir}} = 1 - x_m \tag{1.10.6}$$

Since the summation in (1.10.6) is carried out in respect of r, x_i may be taken from below the sum sign

$$x_i = \frac{1-x_m}{\sum\limits_{\substack{r=1\\r\neq m}}^{n}\frac{1}{k_{ir}}} = (1-x_m)\sum_{\substack{r=1\\r\neq m}}^{n} k_{ir} \tag{1.10.7}$$

Taking the logarithm of the resultant equation and using the derivative:

$$\left(\frac{\partial \ln x_i}{\partial x_m}\right)_{k_{ir}} = \frac{1}{x_m - 1} \tag{1.10.8}$$

Evidently, at $i = m$:

$$\left(\frac{\partial \ln x_m}{\partial x_m}\right)_{k_{ir}} = \frac{1}{x_m} \tag{1.10.9}$$

Thus, the partial derivatives in the expressions (1.10.3) and (1.10.4) are determined by the equations (1.10.5).

Substituting (1.10.5) into (1.10.3) and (1.10.4) and solving the result in systems of linear equations, we obtain the following expressions for the adsorption of components on different dividing surfaces of the ideal multicomponent solution:

$$\Gamma_{m(j)} = -\frac{x_m}{x_j}\sum_{\substack{k=1\\k\neq j}}^{n}\frac{x_k(1-x_k)}{RT}\left(\frac{\partial\sigma}{\partial x_k}\right)_{K_{ir}} - \frac{x_m(1-x_m)}{RT}\left(\frac{\partial\sigma}{\partial x_m}\right)_{K_{ir}} \tag{1.10.10}$$

$$\Gamma_{m(P)} = \frac{-x_m(1-x_m)}{RT}\left(\frac{\partial\sigma}{\partial x_m}\right)_{K_{rj}} - \sum_{i=1}^{n-1}\frac{x_i(1-x_i)}{RT}\left(\frac{\partial\sigma}{\partial x_i}\right)_{K_{rj}}\frac{x_m\sum\limits_{j=1}^{n}P_jY_j - P_ix_nY_m - Y_m\sum\limits_{j=1}^{n-1}P_jx_j}{x_n\sum\limits_{j=1}^{n}P_jY_j} \tag{1.10.11}$$

From the resultant equations we obtain the equations for calculating different variants of determination of adsorption. For example, for an equimolar dividing surface:

$$\Gamma_{m(N)} = \frac{-x_m(1-x_m)}{RT}\left(\frac{\partial\sigma}{\partial x_m}\right)_{K_{rj}} - \frac{Yx_m - Y_m}{Yx_n}\sum_{i=1}^{n-1}\frac{x_i(1-x_i)}{RT}\left(\frac{\partial\sigma}{\partial x_i}\right)_{K_{rj}} \tag{1.10.12}$$

For M- and V-variants

$$\Gamma_{m(V)} = -\frac{x_m(1-x_m)}{RT}\left(\frac{\partial\sigma}{\partial x_m}\right)_{K_{rj}} - \sum_{i=1}^{n-1}\frac{x_i(1-x_i)}{RT}\left(\frac{\partial\sigma}{\partial x_i}\right)_{K_{rj}}\frac{x_m\sum\limits_{j=1}^{n}\overline{V}_jY_j - \overline{V}_ix_nY_m - Y_m\sum\limits_{j=1}^{n-1}\overline{V}_jx_j}{x_n\sum\limits_{j=1}^{n}V_jY_j} \tag{1.10.13}$$

$$\Gamma_{m(M)} = -\frac{x_m(1-x_m)}{RT}\left(\frac{\partial\sigma}{\partial x_m}\right)_{K_{rj}} - \sum_{i=1}^{n-1}\frac{x_i(1-x_i)}{RT}\left(\frac{\partial\sigma}{\partial x_i}\right)_{K_{rj}} \frac{x_m\sum_{j=1}^{n} M_jY_j - M_i x_n Y_m - Y_m\sum_{j=1}^{n-1} M_j x_j}{x_n\sum_{j=1}^{n} M_jY_j} \quad (1.10.14)$$

The expressions (1.10.12)–(1.10.14) may be used to calculate the adsorption in relation to different positions of the dividing surface in two-phase multicomponent systems. When deriving these equations, no assumptions were made regarding the density of the coexisting phases. Consequently, the resultant equations are general. The equations may be used for calculating adsorption at the interface of two non-mixing liquids or at the liquid–vapour interface at temperatures close to critical.

In cases in which it is possible to ignore the content of the components in one of the phases, the resultant equations can be greatly simplified. Actually, at $n_i^\alpha \gg n_i^\beta$ and from (1.10.11):

$$\Gamma_{m(P)} = -\frac{x_m(1-x_m)}{RT}\left(\frac{\partial\sigma}{\partial x_m}\right)_{K_{ir}} + \frac{x_m}{P}\sum_{k=1}^{n} P_k \frac{x_k(1-x_k)}{RT}\left(\frac{\partial\sigma}{\partial x_k}\right)_{K_{ir}} \quad (1.10.15)$$

where $P = \sum_{k=1}^{n} P_k x_k$

For the equimolar dividing surface $P_i = 1; P = \sum_{k=1}^{n} P_k x_k = \sum_{k=1}^{n} x_k = 1$, the equation (1.10.12) has the following simple form:

$$\Gamma_{m(N)} = -\frac{x_m(1-x_m)}{RT}\left(\frac{\partial\sigma}{\partial x_m}\right)_{K_{ir}} + x_m\sum_{k=1}^{n}\frac{x_k(1-x_k)}{RT}\left(\frac{\partial\sigma}{\partial x_k}\right)_{K_{ir}} \quad (1.10.16)$$

It may easily be shown that

$$\sum_{k=1}^{n}\frac{x_k(1-x_k)}{RT}\left(\frac{\partial\sigma}{\partial x_k}\right)_{K_{ir}} = 0 \quad (1.10.17)$$

Here, it is sufficient to derive the expression $\left(\frac{\partial\sigma}{\partial x_k}\right)_{K_{ir}}$, treating the surface tension of the function $\sigma = \sigma(x_1, x_2, \dots, x_n-1)$ and add up in respect to all components taking into account the evident identities: $\sum_{i=1}^{n} x_i = 1$ and $\sum_{i=1}^{n} dx_i = 0$. Consequently equation (1.10.16) with (1.10.17) taking into account has the following form:

$$\Gamma_{m(N)} = -\frac{x_m(1-x_m)}{RT}\left(\frac{\partial\sigma}{\partial x_m}\right)_{K_{ir}} \quad (1.10.18)$$

the equation for adsorption on the equimolecular dividing surface is easily derived directly from the system only in equations (1.10.4), if it is taken into account that $P_i = 1$; $P = 1$ [33]. In fact, in the content of the components in the vapour phase is considerably lower in comparison with the content in the solution $n_i^{\alpha} \ll n_i^{\beta}$, equation (1.10.4) may be presented in the new form:

$$-\left(\frac{\partial \sigma}{\partial x_m}\right)_{K_{i,r}} = \sum_{\substack{i=1 \\ i \neq j}}^{n} \Gamma_{i(N)} \left[\left(\frac{\partial \mu_i}{\partial x_m}\right)_{K_{ir}} + \sum_{\substack{k=1 \\ k \neq j}}^{n} \frac{x_k}{x_j}\left(\frac{\partial \mu_k}{\partial x_m}\right)_{K_{jr}}\right] \tag{1.10.19}$$

Introducing the notations

$$b_m = -\left(\frac{\partial \sigma}{\partial x_m}\right)_{K_{jr}} \quad a_m = \left[\left(\frac{\partial \mu_i}{\partial x_m}\right)_{K_{ir}} + \sum_{\substack{k=1 \\ k \neq j}}^{n} \frac{x_k}{x_j}\left(\frac{\partial \mu_k}{\partial x_m}\right)_{K_{jr}}\right] \tag{1.10.20}$$

Consequently, equation (1.10.20) is written in the form of a system of linear equations:

$$\sum_{i=1}^{n} a_{mi} \Gamma_{i(N)} = b_m \tag{1.10.21}$$

where $m = 1, 2, \ldots, j-1, j+1, n$.

For the adsorption of the m-th component:

$$\Gamma_{m(N)} = \frac{\Delta m}{\Delta} \tag{1.10.22}$$

where Δ is the main determinant of the system of equations, and Δm is the determinant at the unknown $\Gamma_{m(N)}$.

Thus, if the dependence of surface tension and chemical potentials on the composition of the alloy is available, equation (1.10.22) makes it possible to calculate the adsorption according to Guggenheim and Adam (N-variant) in the systems with any number of components. We examine a case of ideal solutions. Therefore, taking into account that $d\mu_i = RTd\ln x_i$ one obtains:

$$a_{mi} = RT\left[\left(\frac{\partial \ln x_i}{\partial x_m}\right)_{K_{ij}} + \sum_{j=1}^{n-1} \frac{x_j}{x_n}\left(\frac{\partial \ln x_i}{\partial x_m}\right)_{K_{ij}}\right] \tag{1.10.23}$$

The partial derivatives, including the last equation, were determined previously and equal to

$$\left(\frac{\partial \ln x_i}{\partial x_m}\right)_{K_{ij}} = \begin{cases} \dfrac{1}{x_m} & \text{for } i = m; \\ \dfrac{1}{1-x_m} & \text{for } i \neq m. \end{cases} \tag{1.10.24}$$

Substituting (1.10.24) into (1.10 .23) gives:

$$a_m = RT\left[\frac{1}{x_m} + \frac{x_m}{x_n}\frac{1}{x_m} + \frac{1}{x_m(1-x_m)}\sum_{\substack{j=1 \\ j\neq m}}^{n-1} x_j\right] = \frac{RT}{x_m(1-x_m)} \tag{1.10.25}$$

at $i = m$.

If $i \neq m$, then

$$a_{mi} = RT\left[\frac{1}{x_m - 1} + \frac{x_m}{x_n} + \sum_{\substack{j=1 \\ j=m}}^{n} \frac{x_j}{x_n(1-x_m)}\right] = 0 \tag{1.10.26}$$

Thus, all the terms of the determinants Δ and Δm, with the exception of the diagonal ones, are equal to zero and it may be assumed that:

$$\Gamma_{m(N)} = \frac{b_m}{a_{mm}} \tag{1.10.27}$$

or, taking equation (1.10.25) and (1.10.26) into account:

$$\Gamma_{m(N)} = -\frac{x_m(1-x_m)}{RT}\left(\frac{\partial \sigma}{\partial x_m}\right)_{K_{ij}} \tag{1.10.28}$$

As indicated by the last equation, the magnitude and sign of the N-variant of adsorption according to Guggenheim and Adam in ideal multicomponent systems, as in a binary systems, is determined by the single derivative $\left(\frac{\partial \sigma}{\partial x_m}\right)_{K_{ij}}$. This derivative characterises the effect of the additions of the m-th component on the surface tension of the multicomponent melt. If the situation with the Gibbs relative adsorption would be the same, it would then be possible to claim unambiguously that this derivative characterises the surface activity of the component in the multicomponent solution. However, this derivative is not linked unambiguously with the relative adsorption.

As indicated by (1.9.6), the sign of relative adsorption is determined by the partial derivative $\left(\frac{\partial\sigma}{\partial x_m}\right)_{X_{k\neq m\neq j}}$. Therefore, the problem of the surface activity of the component in the multicomponent solution requires special attention. The critical analysis of these concepts with reference to the multicomponent systems is very important because the scientific literature contains the erroneous opinion regaraing the complex nature of adsorption processes even in ideal ternary systems. This conclusion is determined by the incorrect definition of the concept of the surface and limiting surface activity of the component in the multicomponent system. Therefore, the critical analysis of these concepts is the subject of a special section.

1.11. Relationship between the values of adsorption in relation to different positions of the dividing surface

As shown previously, depending on the problem to be solved, adsorption is calculated in relation to different positions of the dividing surface. In the case of the multicomponent systems there is the most suitable method of changing the composition for every method of determination of adsorption. For example, if the Gibbs relative adsorption $\Gamma_{m\,(j)}$ is directly linked with the derivative $\left(\frac{\partial\sigma}{\partial x_m}\right)_{X_{i\neq m\neq j}}$, then the adsorption on the equimolar dividing surface $\left(\frac{\partial\sigma}{\partial x_m}\right)_{K_{ij}}$ is the derivative. Therefore, it is interesting to examine the relationship between the values of adsorption in relation to the different positions of the dividing surface. The existence of such a relationship is indicated by the fact that adsorption depends in a linear manner on the position of the dividing surface.

In binary systems, this relationship is easily established on the basis of the equations for calculating the adsorption, derived previously:

$$\Gamma_{1(2)}=\left(1+\frac{P_1\left(n_1^\alpha-n_1^\beta\right)}{P_2\left(n_2^\alpha-n_2^\beta\right)}\right)\Gamma_{1(P)}=-\left(\frac{\partial\sigma}{\partial\mu_1}\right)_T \quad (1.11.1)$$

For the liquid–vapour systems, neglecting the density of the vapour phase, the resultant equation is simpler:

$$\Gamma_{1(2)} = \left(1 + \frac{P_1 x_1}{P_2 x_2}\right)\Gamma_{1(P)} \tag{1.11.2}$$

The relationship (1.11.2), linking together the values of adsorption of the component on different dividing surfaces of the binary solutions, was derived using a different procedure in [46].

In the case of diluted binary solutions $n_i^\alpha \to 0$, equation (1.11.1) and (1.11.2) gives: $\Gamma_{1(2)} = \Gamma_{1(P)}$, *i.e.*, in highly diluted solutions the adsorption is almost completely independent of the position of the dividing surface.

This problem will be examined for the general case of a multicomponent two-phase system with a flat interface. The equations derived for calculating adsorption in the ideal multicomponent systems can be used to determine the relationship between the values of adsorption in relation to different positions of the dividing surface.

Actually, comparing the equations (1.9.6), (1.10.10) and (1.10.11) leads to:

$$\Gamma_{i(j)} = \Gamma_{i(P)} + \frac{\left(n_i^\alpha - n_i^\beta\right)}{\left(n_j^\alpha - n_j^\beta\right)} \sum_{\substack{k=1 \\ k \neq j}}^{n} \frac{P_k}{P_j} \Gamma_{k(P)} \tag{1.11.3}$$

Taking into account the quality (1.4.15), the equation (1.11.3) has the following simpler form:

$$\Gamma_{i(j)} = \Gamma_{i(P)} - \frac{\left(n_i^\alpha - n_i^\beta\right)}{\left(n_j^\alpha - n_j^\beta\right)} \Gamma_{j(P)} \tag{1.11.4}$$

Thus, the relative Gibbs adsorption is linked by means of a simple relationship with the adsorption on any dividing surface whose position is given by the condition $\sum_{i=1}^{n} P_i \Gamma_i = 0$.

Similarly, it is possible to determine the relationship between the adsorption of other dividing surfaces with the relative Gibbs absorption $\Gamma_{i(j)}$:

$$\Gamma_{i(P)} = \Gamma_{i(j)} - \frac{\left(n_i^\alpha - n_i^\beta\right)}{\sum_{k \neq 1}^{n} P_k \left(n_k^\alpha - n_k^\beta\right)} \sum_{k=1}^{n} P_k \Gamma_{k(j)} \left(1 - \delta_{ik}\right) \tag{1.11.5}$$

If it is possible to ignore the content of the component in one of the phases, the equations for adsorption give:

$$\Gamma_{m(P)} = \Gamma_{m(N)} - \frac{x_m}{P}\sum_{k=1}^{n} P_k \Gamma_{k(N)} \tag{1.11.6}$$

$$\Gamma_{m(P)} = \Gamma_{m(r)} - \frac{x_m}{P}\sum_{k=1}^{n} P_k \Gamma_{k(r)}\left(1-\delta_{kr}\right) \tag{1.11.7}$$

$$\Gamma_{m(r)} = \Gamma_{m(N)} - \frac{x_m}{x_r}\Gamma_{r(N)} \tag{1.11.8}$$

$$\Gamma_{m(N)} = \Gamma_{m(r)} - x_m \sum_{i=1}^{n} \Gamma_{i(r)} \tag{1.11.9}$$

$$P = \sum_{i=1}^{n} \mathrm{P}_i x_i \tag{1.11.10}$$

These equations may be used to calculate quite easily the values of adsorption of the components on any dividing surface, if we know the values of the composition of the melt, molar volumes and molar masses of the components, and also their adsorption on one of the dividing surfaces. However, in contrast to the binary systems in which the adsorption of the component of the specific dividing surface is linked only with the adsorption of the same component of other dividing surfaces, in multicomponent systems this relationship is far more complicated.

The equations (1.11.6)–(1.11.9) show that the adsorption of the component on the specific dividing surface of the multicomponent solution is linked with the values of the adsorption of all components on another dividing surface. The values of the relative Gibbs adsorption and of the adsorption on the equimolar dividing surface are linked by the simplest relationship. The equation (1.11.8) shows that the value of adsorption of the m-th component on the dividing surface, corresponding to the zero value of adsorption of r-th component, is fully determined by the values of the more fractions and adsorption of these components on the equimolar dividing surface. Taking into account practical importance, the resultant equations are presented separately for every variant:

$$\Gamma_{m(V)} = \Gamma_{m(N)} - \frac{x_m \sum_{k=1}^{n} \overline{V}_k \Gamma_{k(N)}}{\sum_{i=1}^{n} \Gamma_{i(v)} x_i} \tag{1.11.11}$$

$$\Gamma_{m(M)} = \Gamma_{m(N)} - \frac{x_m \sum_{k=1}^{n} M_k \Gamma_{k(N)}}{\sum_{i=1}^{n} \Gamma_{i(N)} x_i} \qquad (1.11.12)$$

$$\Gamma_{m(V)} = \Gamma_{m(r)} - \frac{x_m \sum_{k=1}^{n} \overline{V}_k \Gamma_{k(r)} (1-\delta_{kr})}{\sum_{i=1}^{n} \Gamma_{i(v)} x_i} \qquad (1.11.13)$$

$$\Gamma_{m(M)} = \Gamma_{m(r)} - \frac{x_m \sum_{k=1}^{n} M_k \Gamma_{k(r)} (1-\delta_{kr})}{\sum_{i=1}^{n} \Gamma_{i(v)} x_i} \qquad (1.11.14)$$

For diluted solutions $x_m \rightarrow 0$ (from 1.11.7) we obtain $\Gamma_{m(P)} = \Gamma_{m(r)}$. Consequently, in the diluted multicomponent solutions, as in binary ones, the adsorption of the diluted substance on different dividing surfaces is the same. This shows that adsorption in this case does not depend on the position of the dividing surface. This behaviour of adsorption is determined by the low density of the number of particles of the dissolved substance in the coexisting phases. In fact, the equation for the dependence of adsorption on the position of the dividing surface (1.3.7) in the limit at $n_i^{\alpha} - n_i^{\beta} \rightarrow 0$ gives $\Gamma_i \approx \Gamma_i^0$, *i.e.* adsorption ceases to depend on the position of the dividing surface.

In conclusion, it should be mentioned that analysis of the Gibbs absorption equation in multicomponent systems was carried out for the first time in the general form. Original equations have been derived for calculating adsorption of components in two-phase multicomponent fluid systems. For systems in which one phase is an ideal multicomponent solution, these expressions may be used to derive explicit expressions for the adsorption of components in relation to the different positions of the dividing surface. It has been shown that the equations for the adsorption variants according to Guggenheim and Adam can be greatly simplified if the content of components in one of the phases is ignored (the ideal multicomponent solution–saturated vapour interface). The equation for the Gibbs relative adsorption does not change in this case.

The resultant equations for the adsorption of components on different dividing surfaces have made it possible to determine in a general form the relationship between the values of adsorption on the different dividing surfaces of the two-phase component system. The expressions (1.11.8) and (1.11.13) were obtained by the authors of [35] by the combined solution of the Gibbs adsorption equation and the Gibbs–Duhem equation. The remaining equations were derived for the first time in this book.

1.12. Adsorption method of determination of the molecular mass of surface-active substances

There are a number of methods of determining the molar mass of surface-active substances, associated with the experimental examination of the physical–chemical properties of diluted solutions: the vapour pressure of the solvent, osmotic pressure, diffusion coefficient, surface tension, *etc*. The methods were described and analysed in [50].

The main shortcoming of these methods is that they can be used only in a specific range of the values of the molar mass. In addition, application of many of these methods requires preliminary calibration on surface-active substances of a specific homologous series. The method based on the measurement of the surface tension of the investigated and reference solution has a number of shortcomings.

Therefore, we developed a more universal method of determining the molar mass of high-molecular compounds based on the phenomenon of adsorption of the component from the diluted solution.

In the methods, the adsorption of the substance from the diluted solution is determined by two independent methods: from the Gibbs adsorption equation, measuring the concentration dependence of surface tension, and using a microtome. In the first case, adsorption is expressed in the numbers of moles per square metre, in the second case, the excess mass of the component in kg/m^2 is determined. These data may be used to calculate the molar mass using the equation derived later in this section.

The mass and, consequently, the mass fraction of the components in the surface layer are determined using a microtome described in [50]. The microtome consists of a sharp blade secured slightly below the surface of the solution.

During rapid rotation of the blade, a layer of the solution is removed from the surface and transferred into a container. Removing the thin layer from the surface, the total mass of the solution, taken from the surface is initially determined. Separating the components, the mass of the components in the removed layer is determined. The solvent may be represented by an easily evaporated substance and the components may be separated by evaporation of the solvent.

The relationship of these quantities with the Gibbs excess energy will be determined. Since the investigated object is the thin layer of the liquid, removed from the surface of the solution, then at the temperatures away from critical temperatures, and ignoring the number of particles in the unit volume of the vapour phase in comparison with the number of the particles in the unit volume of the liquid phase, for adsorption according to Gibbs:

$$\Gamma_i = \frac{N_i - N_i^\alpha}{S} \tag{1.12.1}$$

where N_i is the number of moles of the i-th component in the removed layer; N_i^α is the number of moles of the i-th component required to ensure that the properties of the liquid phase α remain unchanged up to the conventional dividing surface, hypothetically drawn inside the surface layer; S is the total surface area from which the layer is removed.

The adsorption Γ_i, like N_i^α depends on the position of the dividing surface. To solve the given task, it is more convenient to examine the dividing surface defined by the condition

$$\Gamma_i^{(M)} M_1 + \Gamma_2^{(M)} M_2 = 0 \tag{1.12.2}$$

where M_1, M_2 are the molar masses of the components; $\Gamma_i^{(M)}$ is the so-called M-variant of adsorption according to Guggenheim and Adam.

Substituting (1.12.1) into (1.12.2), gives:

$$M_1 \frac{N_1 - N_1^\alpha}{S} - M_2 \frac{N_2 - N_2^\alpha}{S} = 0 \tag{1.12.3}$$

from which $m_1 + m_2 = m_1^\alpha + m_2^\alpha = m_0$, where $m_1^\alpha = M_1 N_1^\alpha$, $m_2^\alpha = M_2 N_2^\alpha$, m_0 in the total mass of the removed layer, m_1 and m_2 are the masses of the components in the removed layer. Consequently, from equation (1.12.1):

$$\Gamma_1^{(M)} = \frac{N_1 - N_1^\alpha}{S} = \frac{1}{M_1}\frac{m_1 - m_1^\alpha}{S} = \frac{1}{M_1}\frac{c_1 - c_1^\alpha}{S} m_0 \tag{1.12.4}$$

where c_1 and c_1^α are the mass fractions of the first component in the removed layer and in the liquid phase, respectively, M_1 is the sought molecular mass.

In the diluted solutions, adsorption $\Gamma_1^{(M)}$ may be determined from the Gibbs equation for the concentration dependence of surface tension:

$$\Gamma_1^{(M)} = -\frac{x_1}{RT\left(1+\frac{x_1 M_1}{(1-x_1)M_2}\right)}\left(\frac{\partial\sigma}{\partial x_1}\right)_T \qquad (1.12.5)$$

where x_1 is the molar fraction of the first component in the solution, σ is the surface tension of the solution.

Since the molar mass of the first component is not available, in equation (1.12.5) it is necessary to transfer from molar to mass fractions. These fractions in the two component solutions linked by the relationship:

$$x_1 = \frac{c_1 M_2}{c_1 M_2 + (1-c_1)M_1} \qquad (1.12.6)$$

Consequently, for the differential dx_1:

$$dx_1 = \frac{M_1 M_2}{\left[M_1(1-c_1)+M_2 c_1\right]^2}dc_1 \qquad (1.12.7)$$

and substituting (1.12.6) and (1.12.7) into (1.12.5) we obtain:

$$\Gamma_1^{(M)} = -\frac{c_1 c_2 (c_1 M_2 + c_2 M_1)}{M_1 RT}\left(\frac{\partial\sigma}{\partial c_1}\right)_T \qquad (1.12.8)$$

Thus, the expressions (1.12.4) and (1.12.8) were derived for $\Gamma_1^{(M)}$. Equating the right-hand part of these expressions, and solving the resultant equations in relation to M_1, gives:

$$M_1 = \left[\frac{1}{c_2}\frac{m_0 RT}{c_1 c_2}\frac{c_1^\alpha - c_1}{S\left(\frac{\partial\sigma}{\partial c_1}\right)} - c_1 M_2\right] \qquad (1.12.9)$$

To determine the molar mass of the component using equation (1.12.9) it is necessary to obtain data on the concentration dependence of the surface tension of the solution and on the composition of the removed layer. Surface tension is measured with high accuracy. Therefore, the accuracy of determination of the molar mass depends almost completely on the improvement of microtome technology or other technical solutions of the problem of determination of the composition of the thin layer, removed from the surface.

The method may be used to determine the molar mass of almost any substance, if the appropriate solvent is selected. The method is free from shortcomings associated with preliminary calibration or the need to use reference samples.

As an example, we carried out experiments with the examination of the molecular mass of phenol using the data for diluted aqueous solutions of phenol.

The experimental data on the surface tension and microtome analysis, required for calculations, were taken from [52]. The surface layer of the aqueous solution of phenol with the thickness smaller than 0.1 mm was separated using a sharper blade moving at a speed of 10.6 m/s, and collected in a small cylinder, coated with silver. The solution was weighed and its concentration was compared with the volume concentration of the solution using an interferometer. The area of the removed layer was 0.031 m^2, and the solution temperature 293 K.

The calculation of the molar mass of phenol using these data gave the value of 100 which coincides with the tabulated data within the range of the error of determination of the phenol concentration in the removed layer.

Thus, the proposed method may be used to determine the molar mass of a large number of substances without any need to use reference samples or preliminary calibration of the device.

1.13. Criteria of the surface activity of a component in multicomponent solutions

The surface-active substances are used widely in the solution of many technical problems [27) and it is not therefore surprising that special attention has been paid to examination of their properties.

To solve a number of practical problems, it would be necessary to carry out at least quantitative evaluation of the surface activity of the component in the solution. Various criteria have been proposed in the literature for the evaluation of the surface activity of elements in binary systems on the basis of other physical–chemical propertie [6, 39, 40]. However, none of these criteria may be regarded as universal and suitable for predicting the capillary activity of components in multicomponent systems. The problem is greatly aggravated by the fact that in multicomponent systems, as shown later, there is ambiguity even in the determination of the surface and limiting surface activities of the component.

In the bad assistance, the sign of adsorption on any dividing surface and, consequently, the distribution of the atoms of the surface between the phases and the surface layer is determined by the derivative $\left(\frac{\partial \sigma}{\partial \mu_m}\right)_T$ which in the case of ideal solutions has the form $\left(\frac{\partial \sigma}{\partial x_m}\right)_T$. On the other hand, this derivative characterises the effect of the dissolved substance on the surface tension of the solution, i.e., determines the capillary activity of the component. Thus, the value and sign of the derivative provide valuable information on the behaviour of the dissolved substance in the solution. This derivative is referred to as the surface activity of the component. The limiting surface activity of the component is the limit

$$G_{0m} = -\lim_{x_m \to 0}\left(\frac{\partial \sigma}{\partial x_m}\right)_T$$

Many investigators have studied the surface activity of the component in binary systems. Various criteria on the surface activity of components in binary systems have been developed by well-known scientists, such as Rebinder [26], Semenchenko [6], Zadumkin [295], Zhukhovitskii [80], Eremenko [297] and Popel' [298]. However, these criteria have been developed for binary systems. No studies have been published to examine the suitability of these criteria for predicting the capillary activity of components in multicomponent systems.

A specific feature of the multicomponent systems, in contrast to bad assistance, is that they contain a large number of methods of variation of the composition. The equations, linking adsorption with the variation of surface tension, depend on the selected method of variation of the composition. As shown previously, depending on the selection of the dividing surface, the adsorption is determined by the derivatives $\left(\frac{\partial \sigma}{\partial x_m}\right)_{K_{ij}}$ (N-variant according to Guggenheim and Adam) or $\left(\frac{\partial \sigma}{\partial x_m}\right)_{x_{j \neq m \neq r}}$ (relative Gibbs adsorption). This situation leads to ambiguity in the determination of the surface activity of late component in the multicomponent systems and this complicates the analysis of the processes of surface separation of substances [8], the phenomenon of concentration buffering of surface tension [6, 13, 47] and also the development of criteria for the surface activity of the component.

The surface activity of a component according to the physical meaning of the word should characterised the distribution of its atoms between the surface layer and the coexisting phases. Adsorption on the equimolar surface (*N*-variant proposed by Guggenheim and Adam) is linked directly with the difference of the concentration the component in the surface layer and in the volume phase. Consequently, the concept of surface activity should be linked with the adsorption in relation to the surface, whose size and sign determined by the partial derivative $\left(\frac{\partial \sigma}{\partial x_m}\right)_{K_{ij}}$. The latter characterises the effect of the addition of the *m*-th component on the surface tension of the solution, and by analogy with the binary systems this derivative may be referred to as the surface activity of the component in the multicomponent solution. Consequently, for the limiting surface activity of the component in their multicomponent solution:

$$G_{0m} = -\lim_{x_m \to 0}\left(\frac{\partial \sigma}{\partial x_m}\right)_{K_{ir}} \qquad (1.13.1)$$

The partial derivative $\left(\frac{\partial \sigma}{\partial x_m}\right)_{x_{j \neq m \neq r}}$, which is directly linked with the relative Gibbs adsorption, was explained efficiently by Popel' [24] who referred to it as the surface activity of the *m*-th component on the *r*-th component in the presence of other components. Of course, the molar fractions of all other components are constant. In the limit at $x_{j \neq m \neq r} = 0$, this derivative characterises the surface activity of the *m*-th component in the binary system *r*–*m*. In this interpretation, this derivative has a specific physical meaning, *i.e.*, it makes it possible to examine the effect of the presence of the third or several components on the capillary activity of the component in the binary system.

Thus, the partial derivative $\left(\frac{\partial \sigma}{\partial x_m}\right)_{K_{ij}}$ characterises the effect of small additions of m-th component on the surface tension of the multicomponent solution. On the other hand, this derivative makes it possible to determine whether the given component is present in the surface layer in an excess amount or whether there is a shortage of this element, in comparison with its content in the solution. Actually, if the derivative has the negative sign:

$\left(\left(\frac{\partial\sigma}{\partial x_m}\right)_{k_{ij}} \prec 0\right)$, then this indicates the excess of the substance in the surface layer ($\Gamma_{i(j)} > 0$, $x_i^\sigma > x_i$, $x_i^\sigma x_i$ are the molar fractions of the component in the surface layer and the solution, respectively), and this component is referred to as surface-active. If the derivative is positive: $\left(\left(\frac{\partial\sigma}{\partial x_m}\right)_{k_{ij}} \succ 0\right)$, the situation is reversed, *i.e.*, the amount of the component in the surface layer is smaller in comparison with the volume of the solution, and the substance is referred to as inactive. Consequently, by analogy with binary assistance, this derivative should be referred to as surface activity, and the limit (1.80) as the limiting surface activity of the m-th component in the multicomponent system.

As mentioned previously, several criteria have been developed for the surface activity of the component in binary systems. It is interesting to examine the possibility of using these criteria for multicomponent systems. According to Rebinder's criterion, of the two components of a solution the active component is the one whose melting point is lower. This criterion is justified only for a specific range of the binary systems and, in principle, cannot be used for multicomponent systems. Semenchenko [6] proposed to use the difference of generalised moments of the solution and the dissolved substance. Zadumkin [82] improve the criterion by introducing the concept of the statistical generalised moment. However, this method also can be used for multicomponent systems. According to the author of [296], the surface active component is the one with a larger molar volume with a higher standard entropy. Nizhenko and Eremenko [297] proposed to determine the surface activity of the component on the basis of the heat of sublimation at the melting point. The equations of the isotherms of surface tension, derived by Zhukovitskii, indicate that the surface activity is determined by the values of the surface tension of the components. A more general criterion for surface activity has been proposed by Popel' [298] on the basis of a simulation thermodynamic equation. However, as reported by Popel', being more general than others, these criteria require data on the activity of the component in the solution. This circumstance complicates the application of these criteria in multicomponent systems for which there are almost no data on the coefficients of thermodynamic activities.

The criterion proposed by Zhukovitskii [80] is very simple and no additional data are required when using this criterion. The equation of isotherms of surface tension proposed by Zhukovitskii shows that in binary solutions, the surface activity of the component is determined by the ratio of the surface tension of the components. This criterion can be used without any special problems for multicomponent systems. One of the components is represented by a pseudocomponent, *i.e.* the multicomponent solution. In the case of ternary systems, a third m-th component is added to the i–j binary solution in the case of variation of the composition of the beam section. In this case, the first component is represented by the i–j binary solution, and the m-th component is added. If the surface tension of the added m-th component is lower than that of the i–j binary alloy, the component reduces the surface tension of the solution and is surface-active. Its content in the surface layer is higher than that in the volume of the solution. On the other hand, if the surface tension of the added m-th component is higher in comparison with that of the i–j binary alloy, the component increases the surface tension of the solution and it is inactive. Its content in the surface layer will be lower in comparison with the content in the volume of the solution. When the values of the surface tension of the added component and of the binary alloy coincide, the surface tension of the solution is independent of the content of the third component. In this case, concentration buffering of surface tension is observed.

Thus, the criterion proposed by Zhukovitskii is the simplest of all examined criteria of surface activity and may be used for multicomponent systems without additional data. According to this criterion, the surface active component in the solution is the one which has a lower surface tension at the given temperature.

In the case of a multicomponent system, the limiting surface activity is determined by the equation (1.13.1). The derivative in this equation is taken with respect to the variation of the composition in the beam section, *i.e.* the m-th component is added to the multicomponent alloy. Therefore, according to Zhukovitskii's criterion, the surface activity in the multicomponent solutions is determined by the ratio of the values of the surface tension of the multicomponent solution and of the added m-th component. The added component should be surface-active, it is surface tension (or molar surface tension) is lower than that of the solution to which it is added.

On the basis of the available experimental data for the concentration dependence of surface tension of nine ternary systems, we verified the reliability of predicting the behaviour of a component in ternary solutions using Zhukovitskii' s criterion. Comparison of the expected behaviour of the third component using this criterion with the experimental data showed differences only in one mercury–lead–bismuth ternary amalgam system. In other systems, the predicted variation of the surface tension is in agreement with the experimental curves.

2

Thermodynamic description of the surface properties by the method of the finite thickness layer

The thermodynamic theory of surface phenomena, developed by Gibbs, is not based on any assumption regarding the thickness and the associated properties of the surface layer.

The second method in which the assumptions on the real transition zone between phases were utilised was developed by Dutch physicists [16, 17], and as the thermodynamic theory of flat surface layers of finite thickness, the method was developed by Guggenheim [18]. The method was developed further, with application to the distortion of the surface, by Rusanov [19, 20]. Various generalisations of the dynamic theory of the surface phenomena were also published by the authors of [21, 40, 54–56].

The specific parameters of the finite thickness surface layer: the molar and partial–molar surface, the composition of the surface layer and its effective thickness. The literature does not contain united views regarding the accurate determination of the thickness of the surface layer and this leads to different approaches to the evaluation of this parameter. A significant contribution to the development of surface theory has been the 'symbaticity criterion', proposed by Rusanov, which may be used for estimating the minimum possible thickness of the surface layer. The application of this criterion made it possible for the first time to prove the thermodynamic insolvency of the monomolecular model of the surface layer for a large number of binary systems [57, 59–62]. However, the 'symbaticity criterion', being the analogue of

Konovalov's law, is not fulfilled when the number of components is greater than two. Therefore, it cannot be used for calculations in multicomponent solutions.

The thermodynamic equations for calculating the composition of the surface layer of two- and three-component solutions have been proposed in [19]. These equations were developed further in [58–63]. The authors of these studies developed a method of calculating the molar surface, composition and thickness of the surface layer of binary systems using the data on the concentration dependence of surface tension and molar volumes. However, the proposed methods are also not suitable for calculating the parameters of the surface layer of multicomponent solutions.

Thus, if reliable methods of calculating the main thermodynamic parameters of the surface layer have been developed for binary system, the solution of the problem for solutions with a large number of components is still in the initial stage. The absence of these methods does not make it possible to use the thermodynamic equations of the method of the finite layer thickness for analysis of the physical processes, taking place on the surface of the multicomponent solution.

In this chapter, several assumptions and considerations of the method of the finance thickness layer are developed further. The first section reviews and analyses comparatively the currently available approaches to the determination of the effective thickness of the surface layer. For pure substances, in the framework of the model of the homogeneous layer, an equation is obtained establishing the relationship of the effective thickness of the surface layer with the surface tension and isothermal compressibility of the liquid.

The relationships which may be used to calculate the molar surface, composition and effective thickness of the surface layer, have been derived for the first time for the multicomponent solution–saturated vapour equilibrium two-phase system.

In the final section, attention is given to various aspects of calculating the partial–molar quantities in multicomponent systems. The equation for the partial–molar surface of the components is made more accurate and calculations are carried out for several metallic melts.

2.1. Effective thickness of the surface layer of solutions

A direct object for examining surface physics is the zone of continuous variation of the local properties situated between the coexisting phases that is anisotropic and heterogeneous in one direction. The zone is referred to as the surface or interphase layer. Because of the small radius of the effect of molecular forces, the properties of the surface layer may greatly change over the distances of the order of intermolecular distances, resulting in steep gradients of the properties and forming a suitable basis for making assumptions on the sudden change (jump) of the properties (in the physical sense of word) at the phase boundary.

The considerations of the surface layer and its thickness greatly changed with time [6, 10]. Starting with Poisson, many investigators assumed that the variation of the properties of the liquid–vapour interface takes place at a certain distance and not suddenly (in jump). However, after publication of Langmuir studies, the considerations of the monomolecular layer became dominant. Evidently, the length of the interphase region depends on the nature of the bordering phases, the radius of the effect of intermolecular forces and also on the parameters of the state of the two-phase system [9–11, 64, 65].

The local properties approached their values in the volume phases asymptotically, which excludes the possibility of formation of a sharp boundary (in the mathematical sense of word) between the surface layer and the homogeneous phase. This resulted in an ambiguity in the definition of the thickness of the surface layer [65, 66].

Often, the definitions of the effective thickness, published by different authors, in the limiting cases ($T \rightarrow 0$ and $T \rightarrow \infty$) result in contradicting data. Therefore, the main approaches to solving this problem will be investigated.

The critical analysis of some of the methods of determination of the thickness of the surface layer was published in [66, 67]. Usually, the thickness of the surface layer is linked with the profiles of local properties (density, concentration, pressure tensor, dielectric permittivity, *etc.*).

However, studies have been published [56, 58, 66, 68] in which this parameter is linked also with the integral properties, such as the average density of the surface layer [58, 63], adsorption and surface tension [56, 66].

The thickness of the surface layers was determined on the basis of the profile the local properties for the first time by the authors of [66]. They proposed two variants. In the first variant, the thickness of the surface layer was determined from the given relative deviation of the local density from its value in the homogeneous phase. The second variant proposes the thickness of the surface layer to be determined on the basis of the intersection of the tangent in the inflection point with the direct lines, corresponding to the values of density in homogeneous phases. This definition was also accepted by Semenchenko [6]. An identical definition was proposed in [70], where the following equation was obtained for the thickness of the surface layer:

$$\tau = \frac{8}{\rho_i^{(\alpha)} - \rho_i^{(\beta)}} \int_0^{\infty} \left[\rho_i^{(\alpha)} - \rho_i(z) \right] dz \tag{2.1.1}$$

The point z=0 corresponds to the value $\rho_i = \frac{\rho_i^{(\alpha)} + \rho_i^{(\beta)}}{2}$.

Equation (2.1.1) and definitions similar to this equation do not provide any information on the thickness of the surface layer. On the other hand, as shown in [66], these definitions are not suitable for all forms of the profile.

According to the authors of [66], the effective thickness of the surface layer may be determined more efficiently as the thickness of the transition zone corresponding to the specific relative difference of some local property. This difference can also be selected so small that the region of the large variation of the given property may be taken into account. The following condition was suggested for the determination of the effective thickness of the surface layer:

$$\left| \frac{L - L^{(\alpha)}}{L^{(\beta)} - L^{(\alpha)}} \right| = C_L, \tag{2.1.2}$$

where L, $L^{(\alpha)}$ is the value of the property at the given point and in the homogeneous phase, respectively; C_L is the sufficiently small constant, specified in advance by the investigator.

The condition (2.1.2) is not suitable for the grain boundaries because in this case $L^{(\alpha)} = L^{(\beta)}$ and the left-hand part tends to infinity. On the other hand, on approaching the critical temperature, the numerator and the denominator tend to zero and the equation (2.1.2) becomes indeterminate.

An identical definition is also used in [71], where the thickness of the surface layer is determined as the distance between the points over which the density of matter in the surface layer increases from $\rho^{(\beta)} + 0.1(\rho^{(\alpha)} - \rho^{(\beta)})$ to $\rho^{(\beta)} + 0.9(\rho^{(\alpha)} - \rho^{(\beta)})$, where $\rho^{(\beta)}$ is the density of the vapour, and $\rho^{(\alpha)}$ is the density of the liquid.

According to the author of [65], the definitions presented previously have one shortcoming. In these definitions, the thickness of the surface layer depends on the parameter specified by the investigator. Evidently, the value of the physical quantity such as the thickness of the surface layer should be independent of the investigator, and should be determined by the nature of the coexisting phases and by physical processes, taking place at the interphase boundary.

Trying to bypass this difficulty, the authors of [65, 67] proposed to use the absolute value of the root-mean-square deviation of the density from its mean value in the given homogeneous phase. Consequently, the following equation may be used for the determination of the effective thickness of the surface layer:

$$\left|\rho^{(\alpha)} - \rho(z)\right| \leq \left|\Delta\bar{\rho}\right|, \tag{2.1.3}$$

where $|\Delta\bar{\rho}|$ is the absolute value of the fluctuation of density in the phase. According to (2.1.3), the homogeneous phase starts where the effects, associated with the presence of the surface, start to be overlapped by statistical fluctuations. However, in addition to the external attractive nature and objectivity, the method has also shortcomings. As is well known, the magnitude of fluctuation depends on temperature, and this is especially evident in the vicinity of the critical point. Consequently, the thickness of the surface layer, determined by the condition (2.1.3) tends to zero at $T \to T_{cr}$ and to infinity at $T \to 0$. This contradicts the experimental facts [69, 72] and existing considerations regarding the surface layer [6, 10, 19].

An original method of determination of the thickness of the surface layer was proposed in [68] for calculating the surface properties of rare-earth metals on the basis of the electronic density profiles. The boundary surfaces between the surface layer and the homogeneous phases, pass through the points, at which the condition

$$n(z) - n^{(\alpha)} = \delta\left(n^{\alpha}\right), \tag{2.1.4}$$

is fulfilled. Here $n(z)$ is the electronic density profile, $n^{(\alpha)}$ is the electronic density in the α-phase, $\delta(n^{(\alpha)})$ is the error of determination of electronic density.

The calculations gave too low values which, according to the authors of [68], was caused by the effects of corrugation of electronic density in the vicinity of the metal–vacuum interface. Regardless of this, the authors of [68] succeeded in the determination of the nature of variation of the thickness of the surface layer in a series of rare-earth metals.

The discussed method, having certain advantages, has also disadvantages mentioned previously. In particular, the accuracy of determination of the physical quantities depends, in addition to other parameters, also on temperature. Consequently, $\delta(n)$ in equation (2.1.4) is not constant and this is clearly evident at temperatures close to critical.

As mentioned previously, the thickness of the surface layer may also be linked with the integral quantities. For example, in [66] the thickness is determined as the width of the zone in which the integral value of the deviation of the local property from its volume value is almost identical with the total integral characteristics of this property, determined in the experiments.

In this aspect, it is necessary to mention the results of [58] in which it was attempted to determine the effective thickness from the average density of the surface layer. The authors derived equations and used them to calculate the thickness of the surface layer of several low-melting metals. It was assumed that the mean density of the surface layer equals 95% of the density of the liquid phase. A shortcoming of the method is that the thickness of the surface layer depends on the parameter given apriori. It is not justified to assume that the ratio of the densities of the surface layer and the liquid phase is the same for all metals. On the other hand, although the authors calculate the temperature dependence of the thickness of the surface layer, it is evident that the ratio of the densities depends on temperature and, at the critical point, these values are equal to each other. Therefore, a critical approach should be used to the temperature dependence of the thickness of the surface layer, calculated in [58].

The comparative analysis of the existing methods of determination of the effective thickness of the surface layer shows that in the given stage they cannot be separated and none of them can be preferred. The selection of a method should be determined by the efficiency of solving the given task. It is therefore necessary to improve further the existing and develop new methods of determination of the thickness of the surface layer. The importance of the task also increases because of the rapid development of experimental methods of surface analysis [73, 76].

2.2. Temperature dependence of the thickness of the surface layer of liquid metals

When examining the surface phenomena by the method of the finite thickness layer, specific parameters are the effective thickness of the surface layer and the associated surface area per 1 mole of matter in the surface layer. For the molar area, the following equation was derived in chapter 1:

$$\omega = \frac{1}{n} N_0^{\frac{1}{3}} \left(\frac{M}{\rho^{\omega}} \right)^{\frac{2}{3}} \tag{2.2.1}$$

where N_0 is the Avogadro number, n is the thickness of the surface layer expressed by the number of monolayers, forming the surface layer; M is the mass of a single mole; ρ^{ω} is the average density of matter in the surface layer.

As mentioned previously, equations were derived in [58] linking together the thickness and mean density of the surface layer. The authors made assumptions which reduced the width of the temperature range in which these equations are valid, and on the other hand, the rigid link of the density of the surface layer with the liquid phase prevents the determination of the temperature dependence of the thickness of the surface layer. Therefore, we attempted to expand the temperature range in which these equations are valid.

Let a one-component liquid be in thermodynamic equilibrium with the natural vapour. Treating the surface layer as an individual phase, the following equation may be written for each phase and the surface layer:

$$\left.\begin{aligned} S^{\alpha} dT - V^{\alpha} dp + N^{\alpha} d\mu &= 0 \\ S^{\beta} dT - V^{\beta} dp + N^{\beta} d\mu &= 0 \\ S^{\omega} dT - V^{\omega} dp + N^{\omega} d\mu &= -A d\sigma \end{aligned}\right\}, \tag{2.2.2}$$

where the indexes (α), (β) and (ω) relate to the liquid, vapour and surface phases, respectively, A is the interfacial surface area. Introducing molar quantities and excluding $d\mu$ and dP from the system (2.2.2) gives:

$$\omega \frac{d\sigma}{dT} = \left(s^{\alpha} - s^{\alpha} \right) - \frac{\upsilon^{\alpha} - \upsilon^{\omega}}{\upsilon^{\beta} - \upsilon^{\alpha}} \left(s^{\beta} - s^{\alpha} \right) \tag{2.2.3}$$

In this equation, all extensive quantities are related to the mole of the substance.

Using the Clausius–Clayperon equation:

$$\frac{dP}{dT}=\frac{s^{\beta}-s^{\alpha}}{\upsilon^{\beta}-\upsilon^{\alpha}}$$

equation (2.2.3) gives:

$$\omega\frac{d\sigma}{dT}=\left(S^{\alpha}-S^{\omega}\right)-M\frac{\rho^{\omega}-\rho^{\alpha}}{\rho^{\alpha}\cdot\rho^{\omega}}\left(\frac{dP}{dT}\right)_{V} \quad (2.2.4)$$

Equation (2.2.4) may be used for evaluating the molar area and, consequently, the thinkness of the surface layer. For this purpose, it is assumed that the surface layer may be produced by isothermal expansion of the liquid to the state with molar entropy S^{ω}, *i.e.*

$$S^{\omega}=S^{(\alpha)}+\int_{S^{\alpha}}^{S^{\omega}}dS=S^{\alpha}+\int_{\rho^{\alpha}}^{\rho^{\omega}}\left(\frac{\partial S}{\partial\rho}\right)_{T}d\rho \quad (2.2.5)$$

Presenting the integral in the form of a series, and using the first approximation, because of the smallness of ρ^{α}–ρ^{ω}, we obtain:

$$S^{\omega}=S^{\alpha}+\left(\frac{\partial S}{\partial\rho}\right)_{T}\left(\rho^{\omega}-\rho^{\alpha}\right) \quad (2.2.6)$$

Taking equation (2.2.6) into account, equation (2.2.4) has the following form:

$$\omega\frac{d\sigma}{dT}=-\left[\left(\frac{\partial S}{\partial\rho}\right)_{T}+\frac{M}{\rho^{\alpha}-\rho^{\omega}}\left(\frac{dP}{dT}\right)_{v}\right]\left(\rho^{\omega}-\rho^{\alpha}\right) \quad (2.2.7)$$

This equation includes the density of the surface layer whose value depends on the selection of the surface model. We examined several general considerations regarding the evaluation of the value of the mean density of the surface layer of the liquid phase which is in equilibrium with the natural vapour.

Evidently, away from the critical temperature the main contribution to the mean density of the surface layer comes from the liquid phase. This is indirectly also indicated by the fact that up to the critical temperature there is a distinctive (optically mirror) surface between the liquid and the vapour. Consequently, it is logical to find a relationship between these quantities.

Experimental investigations have shown that the evidence of the density of the liquid phase on temperature in the vicinity of the critical point may be described by the following equation [77):

$$\left(\rho^{\alpha}-\rho_{K}\right)=a\left(T-T_{K}\right)^{\beta} \tag{2.2.8}$$

Since at the critical point $\rho^{\omega}=\rho^{\alpha}=\rho_{K}$, for the temperature dependence of the density of the surface layer it may also be written:

$$\left(\rho^{\omega}-\rho_{K}\right)=a^{\omega}\left(T-T_{K}\right)^{\beta} \tag{2.2.9}$$

where β is a critical parameter was value for many substances varies in the range from 0.34 to 0.37 [78), and a and a^{ω} are constant coefficients.

After several transformations, from (2.2.8) and (2.2.9):

$$\frac{\rho^{\omega}}{\rho^{\alpha}}=\left(\frac{a}{a^{\omega}}\right)\left(1-\frac{\rho_{K}}{\rho^{\alpha}}\right)+\frac{\rho_{K}}{\rho^{\alpha}} \tag{2.2.10}$$

Evidently, at $T = T_m$, $\rho^{\alpha}=\rho^{\alpha}_{m}$ and $\rho^{\omega}=\rho^{\omega}_{m}$.

Consequently, from (2.2.10):

$$\left(\frac{a}{a^{\omega}}\right)=\frac{K\rho_{d}^{\alpha}-\rho_{K}}{\rho_{d}^{\alpha}-\rho_{K}} \tag{2.2.11}$$

since $K=\frac{\rho_{d}^{\omega}}{\rho_{d}}$.

Taking into account (2.2.11), equation (2.2.10) has a new form:

$$\rho^{\omega}=\rho_{0}+\alpha\rho^{\alpha} \tag{2.2.12}$$

where

$$\rho_{0}=(1-K)\frac{\rho_{d}^{\alpha}\rho_{K}}{\rho_{d}^{\alpha}-\rho_{K}} \quad \text{and} \quad \alpha=\frac{K\rho_{d}^{\alpha}-\rho_{K}}{\rho_{d}^{\alpha}\rho_{K}} \tag{2.2.13}$$

equation (2.2.12) shows that there is a linear relationship between the densities of the surface layer and the liquid phase. Consequently, the mean density of the surface layer may be estimated on the basis of the density of the liquid phase. However, for this purpose it is necessary to introduce the modelling parameter $K=\frac{\rho_{d}^{\omega}}{\rho_{d}}$, whose value is given in advance, or is estimated on the basis of the indirect data.

The equations (2.2.1) and (2.2.7) after simple transformations give:

$$n = -\frac{\dfrac{d\sigma}{dT} N_0^{\frac{1}{3}} \left(\dfrac{M}{\rho^{\omega}} \right)}{\left[\left(\dfrac{\partial S}{\partial \rho} \right)_T + \dfrac{M}{\rho^{\alpha} \rho^{\omega}} \left(\dfrac{\partial P}{\partial T} \right)_{\upsilon} \right] \left(\rho^{\omega} - \rho^{\alpha} \right)} \tag{2.2.14}$$

Formula (2.2.14) may be used to determine the temperature dependence of the thickness of the surface layer of liquid metals. The mean density of the surface layer is determined from equations (2.2.12), (2.2.13). The value of the mean density of the surface layer at the melting point required for these calculations may be determined only for the specific model of the surface layer. The quantitative characteristic of the selected model of the surface layer is the modelling parameter K.

The value of the mean density of the surface layer may be selected only in a specific range. On the one hand, at a negative temperature coefficient of surface tension, the density of matter in the surface layer is lower in comparison with the liquid phase. Another boundary of the range is determined by Rusanov's criterion of the symbatic variation of the composition of the surface layer and the volume phases in the binary systems.

Equation (2.2.14) is the entropy derivative with respect to the density $\left(\frac{\partial S}{\partial \rho} \right)_T$. This derivative may be expressed by the experimentally measured physical quantities. As an example, three identical equations, which may be used depending on the available set of the experimental data, are presented below:

$$\left(\frac{\partial S}{\partial \rho} \right)_T = \frac{\dfrac{\partial \rho}{\partial T} - \beta_T \rho \left(\dfrac{\partial P}{\partial T} \right)_V}{\rho^3 \beta_T} \tag{2.2.15}$$

$$\left(\frac{\partial S}{\partial \rho} \right)_T = \frac{C_P - C_V}{T \left(\dfrac{\partial \rho}{\partial T} \right) - \beta_T \rho \left(\dfrac{\partial P}{\partial T} \right)_V} \tag{2.2.16}$$

$$\left(\frac{\partial S}{\partial \rho} \right)_T = \frac{\dfrac{\partial \rho}{\partial T} - \beta_T \rho \left(\dfrac{\partial P}{\partial T} \right)_V}{\dfrac{T}{C_P} \left(\dfrac{\partial \rho}{\partial T} - \beta_T \rho \left(\dfrac{\partial P}{\partial T} \right) \right) + \rho^3 \beta_S} \tag{2.2.17}$$

where C_P and C_V are the specific heat capacities; β_T and β_S is the isothermal and adiabatic compressibility; $\frac{\partial\rho}{\partial T}$ is the temperature coefficient of density; $\left(\frac{\partial P}{\partial T}\right)_V$ is the temperature coefficient of the pressure of saturated vapours at the given temperature.

These equations determine the relationship between the surface tension and the isothermal compressibility of the liquid. Different ratios of these values were obtained in [9, 68, 79].

However, in contrast to these ratios, the equations mentioned above may be used for calculating the temperature dependence of the thickness of the surface layer. The results of these calculations for the metals examined by the experiments described in this book are discussed in section 4.1.

2.3. Calculation of the composition of the surface layer of multicomponent solutions on the basis of the concentration dependence of surface tension

Rusanov, who developed the method of the finite thickness layer [19, 57], derived thermodynamic equations and inequalities for calculating the main properties of the surface layer of a multicomponent solution, using the data on the concentration dependence of surface tension. However, these equations are differential and problems arise when they are used for calculations in practice. In monographs [19, 20], examination is confined to two- and three-component solutions. As far as we know, the general case of the multicomponent system, described in this paragraph, has not as yet been investigated.

Treating the surface layer as an independent phase and applying the fundamental thermodynamic relationships, the system of equations linking temperature, pressure, surface tension in composition of the coexisting phases has been obtained in the discussed studies:

$$\omega d\sigma = -S_{\alpha\omega}dT + V_{\alpha\omega}dP + \sum_{i,k=1}^{n-1}\left(x_i - x_i^{\omega}\right)g_{ik}dx_k$$

$$\omega d\sigma = \left(S_{\alpha\beta} - S_{\alpha\omega}\right)dT - \left(V_{\alpha\beta} - V_{\alpha\omega}\right)dP + \sum_{i,k=1}^{n-1}\left(x_i - x_i^{\omega}\right)g_{ik}dx_k \quad (2.3.1)$$

where g_k is the second derivative of the Gibbs thermodynamic potential, ω is the molar surface, x_i and x_i^ω are the molar fractions of the i-th component in the solution and in the surface layer, respectively.

The quantities $S_{\alpha\beta}$, $V_{\alpha\beta}$, $S_{\alpha\omega}$, $V_{\alpha\omega}$ are the differential molar effects (entropy and volume) for the process whose direction is indicated by the lower indexes. $S_{\alpha\omega}$ and $V_{\alpha\omega}$ are the differential molar entropy and volume effects of the adsorption from the α-phase.

The system of differential equations (2.3.1) is fundamental in analyses of the surface phenomena in the method of the finite thickness layer. From the system, we derive equations for calculating the composition of the surface layer using the data on the concentration dependence of surface tension.

dP is excluded from the system of equations (2.3.1). Assuming that $V_{\alpha\omega} \ll V_{\alpha\beta}$, the following equation is derived for the isothermal process (T = const):

$$\omega d\sigma = -\sum_{i,k=1}^{n-1}\left(x_i^\omega - x_i\right)g_{ik}dx_k \tag{2.3.2}$$

Equation (2.3.2) characterises the dependence of surface tension on the composition of the multicomponent solution and is an analogue of the Gibbs adsorption equation.

For two-component systems, equation (2.3.2) contains one unknown value of the molar fraction of the component in the surface layer which may be easily determined from the concentration dependence of surface tension and thermodynamic activities. An increase of the number of components also increases the number of unknown quantities in equation (2.3.2) and its solution in relation to the mole factions of the component in the surface layer becomes more complicated.

It should be mentioned that this problem is identical with that which was solved in chapter 1 in analysis of the adsorption phenomena in multicomponent systems. Therefore, as in the calculation of the Gibbs excess quantities, it is taken into account that in the n-component system, $n-1$ variables may change independently. Consequently, using the method of the variation of the composition as the function F_k = const and, at the same time, imposing additional conditions on the differential equations (2.3.2), yields a system of linear equations with respect to $(x_i^\omega - x_i)$:

$$\begin{cases} \omega\left(\dfrac{\partial\sigma}{\partial x_1}\right)_{F_1} = -\sum\limits_{i,k=1}^{n-1}\left(x_i^{\omega}-x_i\right)g_{ik}\left(\dfrac{\partial x_k}{\partial x_1}\right)_{F_1} \\ \omega\left(\dfrac{\partial\sigma}{\partial x_2}\right)_{F_2} = -\sum\limits_{i,k=1}^{n-1}\left(x_i^{\omega}-x_i\right)g_{ik}\left(\dfrac{\partial x_k}{\partial x_2}\right)_{F_2} \\ \cdots\cdots\cdots\cdots\cdots\cdots\cdots\cdots \\ \cdots\cdots\cdots\cdots\cdots\cdots\cdots\cdots \\ \omega\left(\dfrac{\partial\sigma}{\partial x_{n-1}}\right)_{F_{n-1}} = -\sum\limits_{i,k=1}^{n-1}\left(x_i^{\omega}-x_i\right)g_{ik}\left(\dfrac{\partial x_k}{\partial x_{n-1}}\right)_{F_{n-1}} \end{cases} \quad (2.3.3)$$

The system (2.3.3) contains the same number of equations and unknown quantities. Consequently, it is solvable with respect to the unknown values of the molar fractions of the component in the surface layer. For these calculations, as in the case of the boundary solutions, we must have data on the concentration dependence of surface tension and chemical potential of the component in the multicomponent system.

It should be mentioned that the system of equations (2.3.3) was derived for a general case in which the method of variation of the composition is given by the function F_K = const. Specific methods of changing the composition of the solution will be examined.

It was shown in chapter 1 that of the following methods are suitable for practical purposes.

1. With the variation of the composition of the multicomponent solution, the ratio of the molar fractions of all components remains constant, with the exception of one, $F_K = K_{ir} = \frac{x_i}{x_r}$ = const, $i,r \neq m$. This method of the variation of the composition corresponds to the addition of the component denoted by index m.

2. The composition changes in such a manner that the molar fractions of all components, with the exception of two, remain unchanged, *i.e.*, in the solutions, the atoms of one component replaced by the atoms of another component $F_K = x_K$ = const, $K \neq j \neq m$.

From the mathematical viewpoint, both methods of the variation of the composition are equivalent and the selection of a specific method is determined by the ease of solving the given task.

The system of equations (2.3.3) was derived assuming that

$V_{\alpha\omega} \ll V_{\alpha\beta}$, which is equivalent to ignoring the content of the substance in one of the phases. Therefore, these equations, in contrast to the equations derived by the Gibbs method, are not suitable for calculations at the boundary of two dense phases.

Although the system of equations (2.3.3) allows calculation of the composition of the surface layer of real multicomponent solutions, it does not provide explicit expression for the molar fractions of the component in the surface layer. We examined these equations in the case of ideal solutions for which the expression ($\mu_i = \mu_{i0} + RT\ ln\ x_i$) holds. In this case:

$$g_{ik} = RT\left[\left(\frac{\partial \ln x_i}{\partial x_k}\right)_{x_j} + \left(\frac{\partial \ln x_n}{\partial x_k}\right)_{x_j}\right] \tag{2.3.4}$$

After simple transformations, from expression (2.3.4) we obtain:

$$g_{ik} = \begin{cases} RT\left(\dfrac{1}{x_k} + \dfrac{1}{x_n}\right) & \text{for} \quad i = k \\ \dfrac{RT}{x_n} & \text{for} \quad i \neq k \end{cases} \tag{2.3.5}$$

Substituting the value of g_{ik} from expression (2.3.2.3) into the system of equations (2.3.5) and solving it in relation to x_m^{ω} (depending on the method of variation of the composition) we obtain the following expressions:

$$x_m^{\omega} = x_m - \omega \frac{x_m(1 - x_m)}{RT}\left(\frac{\partial \sigma}{\partial x_m}\right)_{Kij} \tag{2.3.6}$$

$$x_m^{\omega} = x_m - \omega \frac{x_m}{RT}\left[\left(\frac{\partial \sigma}{\partial x_m}\right)_{x_{k \neq m \neq n}} - \sum_{i=1}^{n-1} x_i \left(\frac{\partial \sigma}{\partial x_i}\right)_{x_{K \neq m \neq n}}\right] \tag{2.3.7}$$

As indicated by the equations (2.3.6) and (2.3.7), the first method of variation of composition results in a simpler equation for the molar fractions of the components in the surface layer. It is interesting to note that equation (2.3.6) coincides with the identical equation for ideal binary systems. If it is taken into account that the same result was obtained previously in examining the N-variant of adsorption according to Guggenheim and Adam, this may indicate the presence of a closer relationship between the isotherms of adsorption of the component on the secant sections of the multicomponent systems and lateral boundary solutions. The resultant equations may be used for determining the relationship between the content of the component surface layer and its adsorption in relation to different positions of the dividing surface. In fact, taking into account the expressions for $\Gamma_{i(n)}$ and $\Gamma_{i(j)}$, from (2.3.6) and (2.3.7):

$$x_m^{\omega} = x_m + \omega\Gamma_m^n$$

$$x_m^{\omega} = x_m + \omega\left[\Gamma_{m(n)} - x_m\sum_{i=1}^{n}\Gamma_{i(n)}\right]$$

where Γ_m^n is the N-variant of adsorption, and $\Gamma_{i(n)}$ is the adsorption of the i-th component in relation to the dividing surface, given by the condition $\Gamma_n = 0$.

When equation (2.3.6) is used for calculations in practice, it is possible to reduce greatly the error of calculations because, in this case, only the value of one derivative is determined by experiments. When using equation (2.3.7) for these purposes, the values of $n-1$ partial derivatives would be required. Evidently, this results in a large increase of the error of determination of the molar fractions which is determined by the total error of determination of the partial derivatives in equation (2.3.7).

However, on the basis of these consideration it is not possible to make any conclusions on the preference of equation (2.3.7). Cases are possible in practice in which it is difficult to determine the values of the derivative $\left(\frac{\partial\sigma}{\partial x_m}\right)_{Kij}$, whereas the values of the derivatives $\left(\frac{\partial\sigma}{\partial x_i}\right)_{x_{k\neq m\neq n}}$ may be known or easy to determine. There are also a large number of cases in which the combined application of the equations (2.3.6) and (2.3.7) is the optimum approach.

Thus, if the dependence of surface tension on the composition of the multicomponent solution is available, the resultant system of the equations (2.3.3) may be used to determine the composition of the surface layer. It is assumed that the values of the coefficient of thermodynamic activities, required for calculating g_{ik} in real solutions, are available. However, in this case, difficulties associated with the need to know the dependence of the molar surface on the composition of the solution arise in both two-component and in multicomponent systems.

As mentioned previously, because of the principal difficulties in the experimental determination of the molar surface, various assumptions have been made regarding this dependence. In the majority of cases, the dependence of the molar area on the composition is assumed to be an addictive function, although it is more logical to assume that the molar area is additive in relation to the average composition of the surface layer.

In [19] in calculating the composition of the surface layer of the binary systems, the deviations of the molar surfaces from the additivity are taken into account on the basis of the volume effect of displacement. The same procedure may also be used in calculating the composition of the surface layer of multicomponent systems. Actually, the concentration dependence of the molar surface may be presented in the form:

$$\omega = \sum_{i=1}^{n} \omega_i x_i + \Delta\omega \tag{2.3.8}$$

Here $\Delta\omega$ characterises the deviation of the molar surfaces from additivity, ω_i is the molar surfaces of the pure components. If it is assumed that the dependence of the molar volume on the composition in the surface layer is the same as in the volume of the melt, then the method of successive approximations (2.3.3) and (2.3.8) may be used to determine the molar fractions of the component in the surface layer. The connection for the deviation of the molar surface from additivity is found from the well-known equation:

$$\Delta\omega = \frac{1}{n} N^{1/3} \Delta V^{2/3} \tag{2.3.9}$$

where ΔV is the deviation of the molar volume from additivity, n is the thickness of the surface layer, expressed in the number of monolayers. However, the thickness of the surface layer is not available in advance and in the calculations using equation (2.3.2.7) certain assumptions are made regarding its value [62, 63]. However, it is carried out with the restriction to the model of the mono molecular surface layer, although other models are available [63].

2.4. The molar surface, the thickness of the surface layer and their dependence on the composition of the multicomponent solution

In the group of the approximate methods of evaluation of the molar surface special attention should be given the methods proposed in [63] where calculations carried out using Butler's equation [40] $\mu_i^\sigma - \mu_i = \sigma \bar{\omega}_i$ and the equations for calculating the composition of the surface layer [62] yielded equations which make it possible, without any additional assumptions, to determine the composition and effective thickness of the surface layer of binary melts. It is interesting to examine this problem for multicomponent solutions.

The previously derived system of equations (2.3.2.1) may be used to calculate the composition of the surface layer of the multicomponent solutions. However, to carry out these calculations, in addition to the values of surface tension and the thermodynamic activities of the components, it is also essential to have data on the dependence of the molar surface on composition. The value of the molar surface is not determined by experiments, and is calculated from the equation

$$\omega = \frac{f}{n} N_0^{1/3} V^{2/3}, \tag{2.4.1}$$

where V is the molar volume of the solution whose value is determined by experiments, f is the packing coefficient which takes into account the difference between the density of the surface layer and the density of the volume phase.

The equation (2.4.1) shows that the molar surface is directly linked with the thickness of the surface layer, whose value is assumed to be equal to unity and independent of composition. Evidently, the thickness of the surface layer, being one of the main thermodynamic parameters of the surface layer, depends on the composition of the solution and determination of this dependence is one of the most important tasks of the theory of surface phenomena.

To solve this task, the system of the equations (2.3.3) should be supplemented by another independent equation which contains molar surfaces. One of these additional equations is the well-known expression, based on Butler's concept of the separation of a term, associated with the surface tension [40], from the chemical potential of the component in the surface layer:

$$\mu_i^\omega - \mu_i = \sigma \bar{\omega}_i \tag{2.4.2}$$

where $\bar{\omega}_i$ is the partial–molar surface of the i-th component.

The thermodynamic substantiation of the expression was the subject of extensive discussions [13,14,20,80–82]. The thermodynamic analysis of this relationship was carried out in [19] where it was shown that the equation holds for a rigid incompressible surface layer.

The introduction of quantity μ_i^ω as an analogue of the chemical potential is justified by the fact that the conventional selection of the standard state and transition to the thermodynamic activities of the component in the surface layer are possible. An important factor in the justification of the equation is also that the equations of the surface tension isotherms, resulting from this equation, efficiently describes the concentration dependence σ of a large number of binary systems.

Using equation (2.4.2), we expressed the partial molar surface of the i-th component $\bar{\omega}_i$, and substitute it into the well-known thermodynamic identity

$$\omega = \sum_{i=1}^{n} \bar{\omega}_i x_i^\omega \tag{2.4.3}$$

where x_i^ω the molar fractions of the i-th component in the surface layer.

This leads to the expression:

$$\omega = \frac{1}{\sigma} \sum_{i=1}^{n} \left(\mu_i^\omega - \mu_i \right) x_i^\omega \tag{2.4.4}$$

which, taking into account the expression for the difference of the chemical potentials $\mu_i^\omega - \mu i = \sigma_i \bar{\omega}_i + RT \ln \frac{a_i^\omega}{a_i}$, has the following form:

$$\omega = \frac{1}{\sigma} \sum_{i=1}^{n} \left(\sigma_i \omega_i + RT \ln \frac{a_i^\omega}{a_i} \right) x_i^\omega \tag{2.4.5}$$

where σ_i and ω_i is the surface tension and molar surfaces of the pure components, a_i^ω and a_i are the thermodynamic activities of the component in the surface layer and the volume phase, respectively.

Substituting the molar surface by the equation (2.4.5), from the equations (2.3.3) after simple transformations we obtain a system of linear equations which, taking into account the method of variation of the composition of the solution, may be presented in the following form:

1. For the first method of variation of the composition, at $F_m = k_{rj} = \text{const}$

$$\frac{1}{x_m} \left(\frac{d \ln \sigma}{d \ln x_m} \right)_{Krj} \sum_{i=1}^{n} \left(\sigma_i \omega_i + RT \ln \frac{a_i^\omega}{a_i} \right) x_i^\omega = -\sum_{i,k=1}^{n-1} \left(x_i^\omega - x_i \right) g_{iK} \left(\frac{\partial x_i}{\partial x_m} \right)_{Krj} \tag{2.4.6}$$

$m = 1, 2, \dots, n-1$.

2. With the variation of the composition along the lines $F_m = x_m = \text{const}, m \neq r \neq j$:

$$\frac{1}{x_m}\left(\frac{d\ln\sigma}{d\ln x_m}\right)_{Xr\neq mj}\sum_{i=1}^{n}\left(\sigma_i\omega_i + RT\ln\frac{a_i^{\omega}}{a_i}\right)x_i^{\omega} = -\sum_{i,k=1}^{n-1}\left(x_i^{\omega} - x_i\right)g_{iK}\left(\frac{\partial x_i}{\partial x_m}\right)_{Xr\neq mj} \tag{2.4.7}$$

In the resultant equations, the unknown quantities are molar fractions and thermodynamic activities of the component in the surface layer whose values can be determined from (2.4.6) or (2.4.7), solving this system of equation by the method of successive approximations. In the zero approximation, the surface layer is assumed to be an ideal solution ($a_i^{\omega} = x_i^{\omega}$) and from (2.4.6) and (2.4.7) we determine the values of the molar fractions of the component in the surface layer. Assuming that the dependence of thermodynamic activity on the composition and the surface layer out in the volume of the solution is identical, from the values of x_i^{ω} it is possible to determine the thermodynamic activity of the component on the surface layer. Substituting the resultant values of a_i^{ω} into (2.4.6) and (2.4.7) we calculate new values of x_i^{ω}, and so on. In practice, it is usually sufficient to perform one or two cycles since further continuation of this procedure does not lead to any significant corrections in the final results.

Thus, the systems of equations (2.4.6) or (2.4.7) may be used to determine the molar fractions of the component in the surface layer, without any assumptions regarding the concentration dependence of the molar surface. Here it is possible to calculate the values of the molar surface and effective thickness of the surface layer and their dependence on the composition of the solution. Actually, knowing the composition of the surface layer, the values of these quantities may be easily found from the equations (2.4.1) and (2.4.5). The molar surfaces of the pure components, included in the equations (2.4.6) and (2.4.7), were determined from the equations proposed in section 2.2.

The resultant equations were used to calculate the values of the parameters and the surface layer of the binary and multicomponent systems. The surface tension and density of these systems were investigated by experiments. The results and discussion are presented in the following chapters.

Quite cumbersome equations were derived previously for real multicomponent solutions. This complicates analysis and comparison with the equations for simpler systems. These equations are greatly simplified if some permissible assumptions are made. Several of these assumptions will be examined.

In the majority of metallic melts, there are small deviations from the ideal state. On the other hand, Zhukovitskii noted [80] that the probability of formation of an ideal surface layer at the boundary between the solid solution and the vapour is higher than the probability of formation of the volume ideal solution, because the atoms on the free surface of the solution have additional degrees of freedom. It follows from here that with the exception of cases determined by the presence of surface-active groups of atoms on the surface, the surface layer should be closer to the ideal state than the volume phase. This assumption is also supported by the results obtained in the calculation of the coefficient of thermodynamic activity in the surface layer [84, 85]. Calculations of a_i^{ω} for a number of binary systems using the data on surface tension show that the coefficient of the thermodynamic activity of the component is closer to unity in the surface layer than in the volume of the solution.

Taking this into account, in the case of the melts with small deviations from the ideal state, it is fully justified to use the assumption on the ideal state of the surface layer. a_i^{ω} is replaced by the molar fractions of the component in the surface layer and this greatly facilitates calculations using equations (2.4.6) and (2.4.7).

In conclusion, we examine the simplest case in which the surface layer and the volume phase both ideal. In this case, both the calculation equations and calculation procedure are greatly simplified. In fact, from equations (2.3.6) and (2.4.5) it is quite easy to obtain a simpler system of equations for calculating the composition of the surface layer of the multicomponent solution:

$$\frac{1}{RT}\left(\frac{d\ln\sigma}{d\ln x_1}\right)_{Krj}\sum_{i=1}^{n}\left(\sigma_i\omega_i + RT\ln\frac{x_i^{\omega}}{x_i}\right)x_i^{\omega} = \frac{x_1 - x_1^{\omega}}{1-x_1}$$

$$\frac{1}{RT}\left(\frac{d\ln\sigma}{d\ln x_2}\right)_{Krj}\sum_{i=1}^{n}\left(\sigma_i\omega_i + RT\ln\frac{x_i^{\omega}}{x_i}\right)x_i^{\omega} = \frac{x_2 - x_2^{\omega}}{1-x_2}$$

$$\cdots\cdots\cdots \qquad (2.4.8)$$

$$\frac{1}{RT}\left(\frac{d\ln\sigma}{d\ln(x_n-1)}\right)_{Krj}\sum_{i=1}^{n}\left(\sigma_i\omega_i + RT\ln\frac{x_i^{\omega}}{x_i}\right)x_i^{\omega} = \frac{x_{n-1} - x_{n-1}^{\omega}}{1-x_{n-1}}$$

Using the values of the molar fractions of the component in the surface layer, determined from (2.4.8), the molar surface and effective thickness are calculated from the equations (2.4.1) and (2.4.5).

Thus, the resultant equations may be used to calculate the parameters such as the molar surface, composition and effective thickness of the surface layer. Calculations can be carried out for both ideal and real solutions.

2.5. Partial–molar quantities in multicomponent solutions. Partial–molar surfaces

The partial–molar quantities play a significant role in the theory of solutions. Analysis of the dependence of these quantities on the composition of the solution provides useful information on the special features of the structure of phases [86, 87]. This was evidently the reason for the special interest paid recently to examination of these quantities [88, 90]. For example, the authors of [88, 89] developed optimum methods of calculating the partial–molar volumes in the binary systems. Special features of the behaviour and determination of the partial–molar properties in three-component solutions have been examined by the authors of [87, 90, 91]. However, regardless of importance for practice, the partial–molar quantities in the multicomponent solutions have been studied insufficiently.

The concept of the partial–molar surface is introduced in examining the surface layer as an autonomous phase of a finite size. If the problems of calculation of the partial–molar properties of the volume phase have been partially solved or are in the stage of solution, the situation with the determination and calculation of the partial–molar area even in the boundary solutions is far worse. The importance of these problems is caused by the fact that many thermodynamic equations include the partial–molar surface [92–94] and the inaccurate determination of the values of the surface reduce the value of the results obtained from these equations.

Taking this into account, in the first part of this section special attention is given to examining the special features of calculation of partial–molar quantities in multicomponent solutions. The second part is concerned with the analysis of the concept of the partial–molar surface and its calculation in binary and multicomponent solutions.

a) determination of the partial–molar quantities in multicomponent solutions

If the dependence of the molar property Φ (volume, entropy, enthalpy, Gibbs energy, *etc.*) on the composition of the binary solution, determined by the molar fractions x_i, is available, then the partial–molar value of the i-th component $\bar{\Phi}_i$ is determined from the equation:

$$\bar{\Phi}_i = \Phi + (1 - x_i)\left(\frac{\partial \Phi}{\partial x_i}\right)_{T,P} \tag{2.5.1}$$

For the molar property of the multicomponent solution:

$$\Phi = \sum_{i=1}^{K} \bar{\Phi}_i x_i \tag{2.5.2}$$

Differentiating equation (2.5.2) and taking into account the Gibbs-Duhem equation leads to:

$$d\Phi = \sum_{i=1}^{K} \bar{\Phi}_i dx_i \tag{2.5.3}$$

The variation of the composition of the two-component solution they be realised by the only method suitable for this purpose (dx_1 = $-dx_2$) and, consequently, from (2.5.3) we easily obtain the expression (2.5.1). There are several methods of changing the composition of multicomponent solutions.

The most suitable method for the calculations is the method of variation of the composition in which the ratio of the molar fractions of all components, except one, remains constant. In the case of a three-component system, it is the variation of composition in the secants of a concentration angle.

For this case, the equation (2.5.3) assumes the following form:

$$\left(\frac{\partial \Phi}{\partial x_m}\right)_{Kij} = \sum_{i=1}^{K} \bar{\Phi}_i \left(\frac{\partial x_i}{dx_m}\right)_{Kij} \tag{2.5.4}$$

It may be shown that

$$\left(\frac{\partial x_i}{\partial x_m}\right)_{Kij} = \begin{cases} \dfrac{x_m}{x_m - 1} & i \neq m \\ 1 & i = m \end{cases} \tag{2.5.5}$$

Substituting the value of the derivative from (2.5.5) into (2.5.4), after simple transformations, we obtain an equation linking the partial–molar value $\bar{\Phi}_m$ with molar Φ.

$$\bar{\Phi}_m = \Phi + (1 - x_m)\left(\frac{\partial \Phi}{\partial x_m}\right)_{Kij} \tag{2.5.6}$$

When examining multicomponent systems, the second method of variation of the composition is often investigated, in which the atoms of one component replaced by the atoms of another component, with the molar fractions of others remaining constant.

If the composition of the solution is varied by this method, then (2.5.6) is replaced by a more complicated equation:

$$\bar{\Phi}_m = \Phi - \sum_{\substack{i=1 \\ i \neq m}}^{K} x_i \left(\frac{\partial \Phi}{\partial x_m}\right)_{Xj \neq i \neq m} \tag{2.5.7}$$

The resultant equation may be used to calculate the partial–molar properties of any component, if the concentration dependence of the molar properties is known. However, if the partial–molar value in equation (2.5.6) is directly linked with one partial derivative, then n–1 derivatives are required for the determination of this value from (2.5.7). At the same time, equation (2.5.6) enables the graphical method to be used for the determination of the partial molar properties. This method is used widely in the binary systems. Actually, using the equation, it may be shown that the value of the partial–molar property of the m-th component $\bar{\Phi}_m$ is equal to the section produced by the tangent at the point $x_m = 1$, to the isotherm of the molar property constructed in the secant section.

Evidently, from one isotherm of the molar property we can determine the value of the partial–molar property of only one component $\bar{\Phi}_m$. To determine $\bar{\Phi}_i$ of the remaining n–1 components, sections should be made in other directions.

For practical purposes, it is useful to examine another possible case in which it is more convenient to determine the derivatives $\left(\frac{\partial \Phi}{\partial x_m}\right)_{Kij}$ for some components, and $\left(\frac{\partial \Phi}{\partial x_i}\right)_{Xj \neq i \neq m}$ for others.

These cases are encountered frequently in the examination of the properties of ternary systems. In this case, we can use the equation

$$\bar{\Phi}_m = \Phi + (1 - x_m)\left(\frac{\partial \Phi}{\partial x_m}\right)_{Kij} + \left(\frac{\partial \Phi}{\partial x_i}\right)_{Xj \neq i \neq m} \tag{2.5.8}$$

which follows from (2.5.6), if the obvious equality

$$\left(\frac{\partial \Phi}{\partial x}\right)_{Xj\neq i\neq m} = \bar{\Phi}_i - \bar{\Phi}_m \qquad (2.5.9)$$

is taken into account.

Thus, the resultant equations may be used for calculating the partial–molar properties of all components, if the dependence of the extensive property on the composition of the multicomponent solution is known.

2.6. Partial–molar surfaces of components

The concept of the partial–molar surface has been introduced to characterise the surface layer treated as a phase of finite thickness. Like the molar surface, this quantity together with temperature and pressure is also associated with various specific properties, such as surface tension, average composition and thickness of the surface layer. The availability of this value makes it possible in many cases to find the dimensions of absorbing articles and, consequently, solve the problem of the form in which the particles are adsorbed (atom, ion, molecule) [91]. At the same time, the problem of the accurate determination of the partial–molar surfaces has not been solved, are the expressions used for these purposes, are approximate [7, 93].

The equations (2.5.6)–(2.5.8) may be used to determine an exact expression for the partial–molar surfaces. For this purpose, the well-known expression for the molar area

$$\omega = \frac{1}{n} N_0^{\frac{1}{3}} \left(V^{\omega}\right)^{\frac{2}{3}} \qquad (2.6.1)$$

is substituted into equation (2.5.6); here n is the thickness of the surface layer expressed in the numbers of monolayers, V^{ω} is the molar volume of the solution in the surface layer. After relatively simple transformations we obtain:

$$\bar{\omega}_m = \frac{1}{n} N^{\frac{1}{3}} \left[\frac{3V^{\omega} + 2\left(1 - x_m\right)\left(\frac{\partial V^{\omega}}{\partial x_m}\right)_{Kij}}{3\left(V^{\omega}\right)^{\frac{1}{3}}} \right] \qquad (2.6.2)$$

Taking into account the equation (2.5.6), the expression (2.6.2) for the partial–molar volume has the final form:

$$\bar{\omega}_m = \frac{1}{n} N^{\frac{1}{3}} \left[\frac{V^{\omega} + 2\bar{V}_m^{\omega}}{3\left(V^{\omega}\right)^{\frac{1}{3}}} \right] \tag{2.6.3}$$

Equation (2.6.3) may be used to determine the partial–molar surface of the component. However, this equation does not coincide with the well-known equation [92, 93]. Therefore, the proof of the general nature of the equation (2.6.6) will be presented.

It may be shown that the equation (2.6.3) is derived directly from the general thermodynamic definition of the partial–molar surface:

$$\bar{\omega}_i = \left(\frac{\partial \Omega}{\partial N_i^{\omega}} \right)_{T,P,n,N_{j \neq i}^{\omega}}, \tag{2.6.4}$$

where N_i^{ω} is the number of the moles of the i-th component surface layer; Ω is the area of the surface layer. Substituting into (2.6.4) the expression

$$\Omega = \frac{1}{n} \left(N^{\omega} N_0\right)^{\frac{1}{3}} \left(V^{\omega}\right)^{\frac{2}{3}} \tag{2.6.5}$$

where V^{ω} is the volume of the surface layer, after simple transformations we obtain the equation (2.6.3).

On the other hand, the relationship (2.6.3), being the expression for the partial–molar property, according to the Euler theorem, should satisfy the identity:

$$\omega = \sum_{i=1}^{K} \bar{\omega}_i x_i^{\omega} \tag{2.6.6}$$

Substitution of $\bar{\omega}_i$ from equation (2.6.3) into (2.6.6) leads to the relationship (2.6.1), *i.e.,* the identity is fulfilled.

Thus, the partial–molar surface is a thermodynamic parameter of the surface layer and, consequently, also the function of the properties of the surface. To calculate the values of $\bar{\omega}_i$ from equation (2.6.3), it is necessary to have the data on the molar and partial–molar volumes in the surface layer. Experimental measurement of these data is associated with serious and, at the present time, insurmountable difficulties. Therefore, when calculating $\bar{\omega}_i$, certain assumptions and allowances are made and although they are not always strictly fulfilled, they do make it possible to transfer in equation (2.6.3) from the properties of the surface layer to the volume.

In particular, in the presence of not too large deviations of the isotherms of the molar volume from the additive straight lines, and also in the case of formation of strong compounds [91], the molar and partial–molar volumes in the surface may be replaced by the values of these quantities in the volume phase:

$$\bar{\omega}_i = \frac{f}{n} N_0^{\frac{1}{3}} \left[\frac{V + 2\bar{V}_i}{3V^{1/3}} \right], \tag{2.6.7}$$

where f is the packing coefficient close to unity which takes into account the difference of the density of the surface layer from the density of the solution; V and $\bar{V}_i$ is the molar and partial–molar volume of the component in the solution, respectively.

If it is assumed that $V^{\omega} = V = \bar{V}_i^{\omega} = \bar{V}_i$, from equation (2.6.3) we obtain a widely used expression for approximate calculation of $\bar{\omega}_i$:

$$\bar{\omega}_i = \frac{f}{n} N_0^{\frac{1}{3}} \bar{V}_i^{\frac{2}{3}} \tag{2.6.8}$$

It may easily be confirmed that the equations (2.6.7) and (2.56) do not satisfy the Euler theorem.

Thus, the equation (2.6.3) is a strict thermodynamic equation for the partial–molar surface. Regardless of the assumptions made, with a small number of exceptions, the reliable results may also be obtained from equation (2.6.7). Unfortunately, this claim is generally not accurate as regards the calculation of using equation (2.6.8). Therefore, these equations will be analysed in greater detail. In particular, if we examine the case not contradicting to the physical meaning when $V^{\omega} = -2\bar{V}_i^{\omega}$, then the equations (2.6.3) and (2.6.8) give qualitatively different results. In fact, in this case, according to the equation (2.6.3) $\bar{\omega}_i = 0$. This means that the partial–molar surface is equal to zero, although the partial–molar volume differs from zero. This conclusion contradicts the equation (2.6.8) which shows that $\bar{\omega}_i = 0$ only at $\bar{V}_i = 0$. This is possible in cases in which the variation of $\bar{V}_i^{\omega}$ takes place only as a result of the increase of the thickness of the surface layer. For a large number of binary systems in calculation of the surface tension using of the currently available theoretical equations [7, 84, 92, 93], the partial–molar surface is determined from the approximate equation (2.6.8). It is interesting to analyse the error made in this case and develop a criterion to explain for which systems it is necessary to improve accuracy and determined systems for which the error in

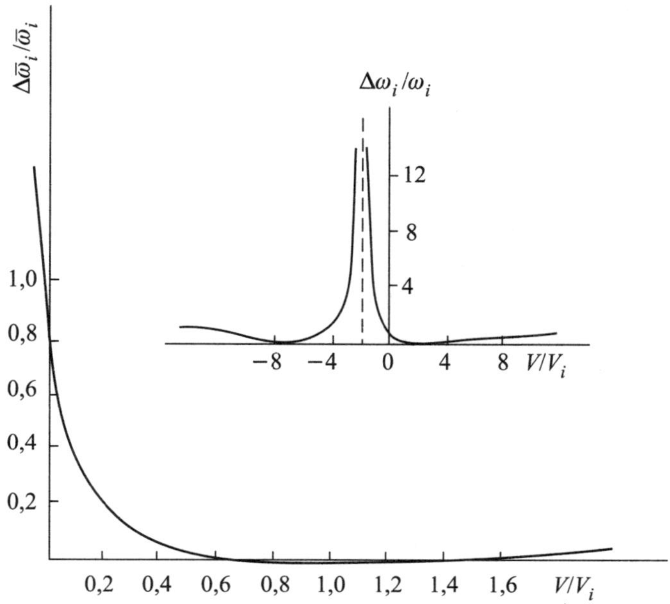

Figure 2.1. Error in calculating the partial–molar surface.

determination of $\bar{\omega}_i$ is in the permissible range.

For this purpose, we introduce a coefficient, characterising the difference of the molar volume of the solution from the partial–molar volume of the i-th component $K=V/\bar{V}_i$. It may easily be shown that the error is fully determined by the value of this coefficient, and at $K = 1$ equation (2.6.3) is transformed into (2.6.8) and the error is equal to zero. In fact, the relative error in the determination of $\bar{\omega}_i$ using equation (2.6.8) is equal to:

$$\left|\frac{\Delta\bar{\omega}_i}{\bar{\omega}_i}\right| = \left|1 - \frac{3K^{\frac{1}{3}}}{K+2}\right| = f(K) \qquad (2.6.9)$$

where $\Delta\bar{\omega}_i$ is the difference of the values of $\bar{\omega}_i$ determined from the equations (2.6.3) and (2.5.6); $f(K)$ is a function characterising the magnitude of the error. The vertical bars in [2.6.9] show that the absolute value of the function is considered. The graph of this function, which may be referred to as the error function, is shown in Fig. 2.1. As indicated by the graph, regardless of its complicated form, the error function has greatly 'diffuse' minima $K = 1$ ($\bar{V}_i = V$) and $K = -8$. Because of these minima in the cases in which K is not higher, the error of determination of $\bar{\omega}_i$ using equation (2.6.8) does not exceed 5%. However, at $K \geq 4$ or $K \leq 0.2$, the error reaches 20%. Of greatest interest is the point $K = -2$ when the function tends to infinity and this point corresponds to the previously examined case in which the equations (2.6.3) and (2.6.8) give qualitatively different results.

Thus, the equations (2.6.3) and (2.6.8) give different results even in the case of additivity of the molar volumes, and this difference is fully determined by the ratio $K = V/\bar{V}_i$. For the majority of binary metallic systems available at the present time, the value of K is in the range which is such that the error in determination $\bar{\omega}_i$ does not exceed the permissible range. However, for every specific system, the problem of the errors should be solved on the basis of the experimental data on the molar and partial–molar volumes.

Since the values of the partial–molar surfaces, determined from equation (2.6.3) and (2.5.6) may greatly differ, it is interesting to compare them for a number of systems and verify the effect of these differences on the behaviour of the theoretical isotherm of surface tension [92].

For this purpose, the data on the densities of five binary systems: cobalt–silicon, nickel–silicon [94], iron–silicon [95], gallium–bismuth [96] and gallium–indium [97], were used to calculate, using equations (2.6.3) and (2.6.8), the partial–molar surfaces of components and the results were subsequently used to determine the values of the surface tension using the Popel'–Pavlov equation [92].

The selected systems are interesting because of the special features of the variation of the partial–molar volumes and, consequently, of the surfaces with composition. In the gallium-bismuth and gallium–indium melts, with a large difference of the values of the molar volume of the components, the dependence of the molar volume on the composition is close to linear. In the remaining three systems, there is a compression accompanied by a large deviation of the molar volumes from the additive straight lines.

The results of calculations of the partial–molar surfaces are presented in Figs. 2.2 and 2.3. In the nickel–silicon system, the difference in the values of $\bar{\omega}_i$ was of the intermediate value between iron–silicon and cobalt–silicon and is not presented here. On the whole, the differences were small but require attention. Large differences were found in the cobalt–silicon system in which at low concentrations of cobalt the differences in the values determined using the discussed equations reached 15%. The isotherms of the partial–molar surfaces, calculated from the equation (2.5.6), have special features not reflected by the equality

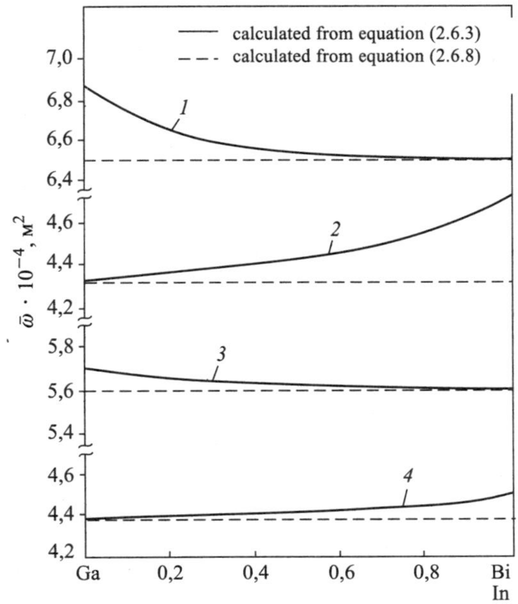

Figure 2.2. Isotherms of partial–molar surfaces of gallium (2), bismuth (1) in the gallium-bismuth system at 573 K, and of gallium (4), indium (3) in gallium–indium melts at 573. Solid line– calculated from equation (2.6.3), dashed line – calculated from equation (2.6.8).

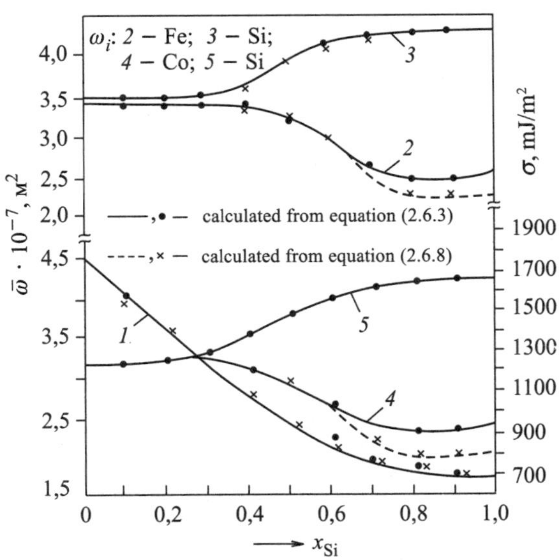

Figure 2.3. Isotherms of surface tension (1) and partial–molar surfaces (2–5) in iron–silicon (2–3) and cobalt–silicon melts (4–5) at 1823 K.

(2.6.3): the non-monotonic nature of variations, the presence of a minimum.

It is interesting to note that the detected differences in the values have only a slight effect on the values of surface tension calculated from equation [92]:

$$\sum_{i=1}^{n} V_i \exp\left(\frac{\sigma - \bar{\sigma}_i}{RT}\right) = 1,$$

where V_i is the volume fraction of the components.

Even in the cobalt–silicon system, the differences in the values are not significant. They do not extend outside the error range of the experimental determination of surface tension. The reasons for the low sensitivity of surface tension to the errors $\Delta\bar{\omega}_i$ may be easily found by determining the error $\Delta\sigma$ from equation (2.6.10). For a binary solution taking into account the Gibbs–Duhem equation, linking the partial–molar surfaces:

$$x_i^{\omega}\Delta\bar{\omega}_1 + x_2^{\omega}\Delta\bar{\omega}_2 = 0 \qquad (2.6.10)$$

from equation (2.6.8) we obtain:

$$A_1 \exp\left(\frac{\sigma}{RT} d\bar{\omega}_1\right) + A_2 \exp\left(-\frac{x_1^{\omega}}{x_2^{\omega}}\frac{\sigma}{RT} d\bar{\omega}_1\right) = 1 \qquad (2.6.11)$$

where

$$A_i = V_1 \exp\left(\frac{\sigma - \bar{\sigma}_i}{RT}\bar{\omega}_i\right)$$

The value $d\bar{\omega}_i$ has the opposite effect on the terms of the sum in the left hand part of the equality (2.6.11); this is due to the opposite signs of the exponents. Therefore, the total effect of $d\bar{\omega}_i$ on equation (2.6.11) and, consequently, on the surface tension is small.

Thus, the isotherms of the partial–molar surfaces, calculated from the equations (2.6.3) and (2.6.8), may greatly differ from each other. However, these differences do not have any significant effect of the values of surface tension, calculated from the simulation–thermodynamic equations.

3

The method of measurement of the surface tension and density of metals and alloys

Because of the rapid development of new areas of science and technology, the requirements on the accuracy of experimental data become more and more stringent. Therefore, special attention is paid in the literature to the problems of improvement of the existing and search for new methods of measuring the surface tension and density of substances in both the liquid and solid state [98–103, 305].

Experimental examination of the surface properties of metals and alloys is associated with considerable difficulties which increase appreciably in the examination of the multicomponent systems. The most important task is to increase the productivity of labour-consuming experiments in examination of the concentration dependence of surface tension and density. Analysis of the literature sources shows that there are two directions of the development of the methods of measuring these properties. The first task is to increase the accuracy of measurements, the second to increase the productivity of experiments.

In the examination of multicomponent systems, it is more efficient to use high-productivity devices which greatly reduce the labour content, the duration of experiments and the consumption of expensive metals. The volume of experimental measurements may be reduced whilst maintaining the required accuracy by using the methods of mathematical experiment design. In this case, the advantages of precision measurements are obvious.

In the examination of multicomponent systems we have improved the method of measurement and design devices for increasing the accuracy of measurements and productivity. Depending on the investigated objects, measurements are taken using two different devices.

In the first device, density iss measured by the pycnometric method and surface tension by the method of maximum pressure in the droplet. The main shortcoming of the method is the fact that the device must be opened to prepare an alloy with a new composition. This is associated with the need for rinsing, drying and long- term vacuum heat treatment. The method is also characterised by a high consumption of expensive metals.

These shortcomings are eliminated in the combined device in which it is possible to prepare alloys of different compositions, without opening the device and disrupting the thermal vacuum conditions, and the temperature dependence of the surface tension and density of these melts can be investigated. Surface tension is measured, as in the first device, by the method of maximum pressure in the droplet, and the density with an improved areometer. The device makes it possible to take into account the effect of capillary forces and thermal expansion of the areometer on the results of density measurements. The device shortens the experimental time 8–10 times and also the consumption of expensive metals.

When examining multicomponent systems using the methods of mathematical experiment design, measurements were taken in the first device. The combined device was used in the remaining cases.

In addition, other devices and attachments have been designed for examination of the properties of non-volatile low-melting (up to 773 K) substances. They are also described in this chapter. It should be mentioned that the experimental investigations, including the development of new devices, have been carried out by the author of the book together with undergraduates and colleagues under the scientific supervision of K.I. Ibragimov. The author is grateful to them for their help and has tried to reflect to the maximum extent their participation in this complicated and time-consuming work, with references made to joint scientific articles.

3.1. Selection of the method of measuring surface tension and density

A critical analysis of the currently available methods of measuring surface tension and density has been published in [6, 7, 22, 98, 104, 105] and in books [106–108]. It was shown in these studies that of the currently available methods of measuring the surface tension of solutions, the theoretically most justified and developed in the apparatus sense of the word are the methods of the droplet on a substrate or suspended on a bar [98, 99, 108–110, 114] and the method of maximum pressure in a gas bubble [120–122] or in a droplet [6, 100]. Regardless of a number of unsolved problems, contactless methods of measuring surface tension and density are also promising [101, 115]. Each of these methods has its advantages and disadvantages and also specific ranges of application. Therefore, the selection of a specific method is determined by tasks and special features (chemical activity, volatility) of the investigated object, and also by the temperature range in which the measurements are taken.

The surface tension of salt and oxide melts is determined in most cases by the method of maximum pressure in a bubble [110]. This is determined by the low sensitivity of the surfaces of these melts to the presence of oxygen in the gas phase. The application of the method for measuring the surface tension of metallic melts is associated with certain difficulties caused by the strong effect of the gas phase on the surface properties of the melt [125, 128].

The method of the sessile droplet (or the method of the 'large' droplet) is used in high temperature investigations and for measuring the surface tension of liquids in the supercooled conditions [129). A significant contribution to the theoretical justification of the method and also development of the measuring procedures has been provided by Eremenko, Ivashchenko, Nizhenko [77], Popel' [13], Zadumkin,, Khokhonov, Alchagirov [141, 306], Khantadze [106, 118], and their students and colleagues. The advantages of the method include the comparatively large possibilities of automation of the measurements and reserves for increasing the accuracy of processing the profiles of the droplet using advanced computers [99, 123–126]. The special features of the application of the method and apparatus for measuring surface tension and density are described in particular in [306].

At experiment temperatures below 773 K it is recommended to

use the method of maximum pressure in the droplet because this method has a number of advantages, with the most important advantage being the high accuracy of measurements. Therefore, the majority of the most reliable measurements of the surface tension of low melting metals and alloys has been carried out by this method.

The extensive application of the method of maximum pressure into practice of precision measurements of surface tension has been facilitated by the so-called gravitational devices, developed by Pugachevich, Pokrovskii, Lazarev, Ibragimov [100] and their colleagues [120, 121, 127]. As shown below, further modifications of these devices may be useful for solving other problems associated with the improvement of the measurements of the surface tension and density at low melting metals.

Taking these considerations into account, we selected the method of maximum pressure in the droplet for measuring the surface tension of low-melting metals (indium, tin, lead, thallium, gallium and bismuth) and of their binary and multicomponent melts. Measurements were taken in devices produced from borosilicate molybdenum glass in the temperature range from the liquidus line to 773 K. The theoretical fundamentals of the method have been analysed in detail in a monograph [6]. Therefore, the content and special features of the application of the method for the examination of metals and alloys will be examined briefly.

The method of maximum pressure in the droplet or gas bubble is based on the fact that we determine the pressure, required for pushing a gas bubble through a capillary with the known radius, immersed in the investigated liquid, or for pushing the droplet of the investigated liquid through the given capillary.

Ignoring cumbersome mathematical considerations, which have been published in [6], attention will be given to the principal assumptions and considerations made in deriving the calculation equation. It has been assumed that the droplet is strictly spherical, the maximum pressure in the droplet is independent of the wetting angle, and the droplet radius is equal to the internal radius of the capillary through which the droplet is pushed. Under these fully acceptable assumptions, an equation was derived linking the interfacial tension with the maximum pressure in the bubble (droplet) at which the droplet collapses:

$$\sigma = \frac{rP_m}{2}\left[1 - \frac{2}{3}\frac{g(\rho_2-\rho_1)}{P_m}r - \frac{1}{6}g^2\frac{(\rho_2-\rho_1)^2}{P_m^2}r^2\right] \tag{3.1.1}$$

where P_m is the maximum pressure in the droplet, r is the radius of the capillary, g is the gravitational acceleration, ρ_1 and ρ_2 are the densities of the contacting liquid phases.

The equation (3.1.1) was derived from the first time by Cantor and Verschaffelt and is therefore often referred to as the Cantor formula. In the case of the liquid–vapour interface, the condition ($\rho_1 \ll \rho_2$) is fulfilled away from the critical temperature, and the equation (3.1.1) may be presented in the following form:

$$\sigma = \frac{r}{2}P_m\left[1 - \frac{2}{3}\frac{g\rho r}{P_m} - \frac{1}{6}\frac{g^2\rho^2}{P_m^2}r^2\right] \tag{3.1.2}$$

The assumptions made when deriving this formula increase the requirements on the capillary used in experiments [6, 121, 130]. On one hand, the assumptions on the strictly spherical form of the droplet and the equality of the droplet radius to the internal radius of the capillary require the capillary to be of a small diameter. On the other hand, a decrease of the size of the capillary increases its ellipticity and the relative error of radius measurements.

The selection of the optimum capillary is a complicated, time- and labour-consuming procedure and determines the reliability of the experimental results. For the devices used in our experiments we selected capillaries whose ellipticity did not exceed 0.5%. For this purpose, approximately 100 capillaries were drawn prior to the experiments. The cross-section of the capillary was examined under a microscope and the capillaries whose section was free from chips, cracks and other damage were selected.

The radius of the capillary was determined using a microscope with the accuracy of 10^{-6} m. If the radius of the capillary and its ellipticity fitted in permissible ranges, multiple measurements of the dimensions of the capillary were taken. Mean values were used in the calculations. The radius of the capillaries used in our experiments varied in the range from $2\cdot10^{-4}$ m and to $3\cdot10^{-4}$ m.

As mentioned previously, an important stage in the improvement of the method of experimental measurement of surface tension was the development, by Pugachevich, of devices working on the gravitational principle. These devices resulted in a revolution in the experimental examination of the surface properties of metals and alloys because they enabled the accuracy of measurement of surface tension to be increased by almost an order of magnitude. A comparative analysis and detailed description of these devices were published in [6, 130].

The most important advantages of the gravitational devices are: efficiency of experiments, the possibility of multiple repetition of measurements, and also of preparation and examination of the surface tension of a large number of alloys of different composition; the possibility of unlimited mixing of the melt; comparatively short measurement time; the possibility of establishing the thermodynamic equilibrium of the liquid with the vapours prior to measurements. These advantages enable a high accuracy of the measurements to be achieved.

However, as mentioned in [100,102,119], a common disadvantage of these devices is that they do not measure the density of the melt required for calculating the surface tension. This makes it necessary to carry out the approximate evaluation of the density values or conduct additional experimental investigations. Separate measurements of the surface tension and density are irrational because in practice it is difficult to prepare alloys with the same concentration in different devices. In addition, the consumption of expensive metals is high and working time long. Finally, it is very important to measure these quantities in the same conditions of thermal vacuum treatment of the device, the purity of the initial components and thermostatic control of the melt.

The task of combined measurement of the surface tension and density has been solved in [119] in which a device combining the Pugachevich gravitational method of measurement of surface tension, a pycnometer for measuring the density and a cylinder for measuring the work function of the electron are described. A device of similar design was used in our experiments in the examination of ternary systems using the methods of mathematical experiment design. The design of the device and experimental procedure are described in section 3.4. In [100] a variant of this device is described in which special features of examination of the melt with volatile components are taken into account. This represents advances in the development of measuring techniques. However, these devices have been designed for examining the temperature dependence of the surface tension and density of a single melt. In examination of the concentration dependence of the surface tension and density of binary and, even more so, multicomponent melts, it is important to examine their properties for a large number of compositions. Therefore, in examination of the multicomponent systems, special attention is given to the problem of increasing the productivity of experimental investigations without any larger decrease in the accuracy of measurements.

This task can be efficiently solved by the development of combined devices in which it is possible to prepare alloys of different concentrations, without opening the device and disrupting thermal vacuum conditions, and measurements of surface tension and density at different temperatures can be taken. The development of these devices frees the experimentator from carrying out labour-intensive investigations, associated with the need for rinsing, drying and long-term thermal vacuum treatment of the device prior to measurement of the properties of individual alloys. On the other hand, it is important to ensure identical thermal vacuum conditions of investigation of the properties of alloys of different concentration.

The task of the combined measurement of the surface tension and density of the melt of different concentrations, prepared in the device, without opening the device and disrupting thermal vacuum conditions, has been solved in our work by the application of special areometers for density measurements. In contrast to the pycnometric method of density measurements, the application of the areometer does not require weighing of the investigated substance and enables one device to be used for investigating tens of melts of different concentration.

3.2. Special features of measurement of the density of liquid metals and alloys with areometers

The majority of methods of surface tension determination is associated with the need to know the density of the investigated liquid. In addition, the density, like viscosity, electrical conductivity, surface tension, is a structure-sensitive characteristic of the melt and the accurate measurement of this parameter is also important for investigating the structure of the liquid. The problems of accurate measurement of the density are important in the development the measuring devices in various areas of science and technology [132]. It is therefore very important to develop new and improve the existing methods of density measurements.

At present, there is a large number of reliable methods of measuring the density of metals and alloys in a wide temperature range [116, 117, 133]. New methods of measurement, based on the utilisation of the dependence of various physical quantities of the density of matter, are being rapidly developed. They include methods based on the variation of: radiation (radioactive, x-ray) passing through matter [134–137]; the speed of propagation of sound [138]; frequency and amplitude of vibrations of an auxiliary body which is in contact with the investigated substance [139], *etc*. Comparative analysis and detailed description of the main methods and devices for density measurements have been published in [117]. Therefore, here we examine only the measurement methods which were used for examination of the density of liquid metals and alloys. These are the areometric and pycnometric methods. Initially, we examine the special procedural features of the application of areometers for the measurement of the temperature dependence of the density of liquid metals and alloys.

The externally simple procedure of the areometric method of determination of density masks considerable difficulties and problems, and the inefficient solution of these problems greatly increases the measurement error. The application of areometers for precision measurements of the density in scientific and research investigations requires taking into account accurately the effect of capillary forces and thermal expansion of the areometer on the readings of the device. On the other hand, stringent requirements are also imposed on the design of the areometer. In particular, the areometer should have the form and dimensions ensuring stable equilibrium when floating. This is possible if the following conditions are fulfilled:

1. The centre of gravity of the areometer and the point of application of the pushing force should be situated on the same vertical straight line;
2. The centre of gravity of the areometer should be located below the point of application of the pushing force.

The unstable equilibrium of the areometer in floating greatly complicates accurate measurement of the density of liquid metals and alloys taken in closed glass devices in high vacuum and elevated temperatures. Therefore, in this case, it is very important to fulfil these conditions in the manufacture of the areometer.

The first condition is fulfilled by designing the areometer in such a manner that its form is symmetric in relation to the vertical axis. The second condition is fulfilled by filling the lower part of the body with ballast. In our experiments, the ballast was represented by cylindrical pieces of tungsten so that areometers with wide ranges of density measurements could be produced.

If these conditions are fulfilled, the areometers may be used for measuring the density of metals and alloys. The areometric method of determination of density has a number of advantages in comparison with other methods, with the main advantage being the absence of the need for weighing the investigated melt. Because of the previously mentioned difficulties, the areometric method is not used for measuring the density of metallic melts.

In [140], Ibragimov *et al* proposed for the first time a device for the combined measurement of surface tension by the method of maximum pressure in the droplet, and measurements of density using the areometer. The development of this device was the start of an important and intensive work on the improvement of devices for the method of combined measurement of the surface tension and density of metallic melts. However, the method for measuring the density, proposed in [140], does not make it possible to take into account the effect of capillary forces on the readings of the areometer and discussion resulted in large errors in individual cases (up to 5%).

The effect of capillary forces was taken into account by the authors of [122, 141]. For this purpose, the chamber of the areometer was fitted with a capillary discharge tube. Consequently, the equations, characterising the equilibrium of the liquid in the areometric chamber and the capillary discharge tube and the conditions of free floating of the areometer, taking into account the evident inequality $R_1 \gg R_2$ (R_1 and R_2 are the internal radii of the cylinder in which the areometer is placed, and of the outlet of the discharge tube), the following equation is obtained for calculating density:

$$\rho = \frac{P_0}{g\left(V_0 + \pi rh\right) - \pi Rr\Delta h} \qquad (3.2.1)$$

In this equation, P_0 is the weight of the areometer in vacuum, V_0 is the total volume of the areometer, r is the radius of the bar, N with the radius of the capillary discharge tube, Δh is the difference of the levels of the liquid in the areometric cylinder and the capillary discharge tube, h is the distance from the level of the liquid to the upper end of the bar of the areometer.

It was shown in [142] that the capillary phenomena in the measurement of the density with the areometer may be taken into account if two areometers are used instead of one. According to the author, the radii of the bars in these areometers should be identical but this is not always fulfilled. It is difficult to use the bars with the same radius, and if the condition of the quality of the radius of the buyers is not strictly fulfilled, errors may form. It may easily be shown that the effect of the capillary forces may be taken into account by two areometers with bars of different diameters. Actually, writing for every areometer the conditions of its free-floating, we obtain:

$$m_1 g + 2\pi r_1 \sigma \cos\theta = \rho g V_1;$$
$$m_2 g + 2\pi r_2 \sigma \cos\theta = \rho g V_2. \qquad (3.2\ .2)$$

The solution of these equations gives the calculation formula

$$\rho = \frac{m_2 - \frac{r_2}{r_1} m_1}{V_2 - \frac{r_2}{r_1} V_1} \qquad (3.2.3)$$

where m_1 and m_2 are the masses of the areometers, r_1 and r_2 are the radii of the bars, V_1 and V_2 the volumes of the parts of the areometers immersed in the liquid.

Thus, when using two areometers, only two quantities V_1 and V_2 are determined in the measurements. The remaining quantities in the equation (3.2.3) are determined in the process of preliminary calibration of the areometers.

It should be mentioned that the calibration of the measuring devices is a more important part of the experimental studies even than the measurements themselves. The accurate and efficient calibration may greatly reduce the systematic error of the experiments, and vice versa. Regardless of this, far less attention has been paid to the problems of calibration in the literature than to the measurement process. It may be possible that this is the cause of the difference of the experimental data of the values of the density of pure metals [143], obtained by different authors. In many cases, these differences greatly exceed the total error of measurement obtained by different authors. Therefore, the special features of calibration of the areometers will be discussed in greater detail.

As shown previously, calibration is the preliminary determination of the physical quantities included in the equation for the calculation of density: mass, the radius of the bar, and the volume of the part of the areometer immersed to a specific depth. The mass and the radius of the bar are measured by accurate devices and do not change during service. Analysis of the total error of measurement shows that the main part of the error of measurement of the density by the areometer is caused by the inaccurate determination of the volume of the part of the areometer, immersed in the liquid. Therefore, calibration of the areometer is reduced to the determination of the dependence of this quantity on the depth of immersion of the areometer which is counted from the upper edge of the bar, from the mark made in advance on the bar.

The following methods are used for calibrating the areometers: a) the hydrostatic weighing method, b) calibration on pure metals with the known density.

A common disadvantage of these methods is that it is necessary to take into account the capillary forces also in calibration and this results in an additional error. On the other hand, in calibration on pure metals an additional error is made as a result of the inaccurate data on the density of these metals.

At present, only mercury may be used as a reference metal of all the metals available. The density of mercury has been measured with high accuracy [144]. However, in contrast to the pycnometers, the areometers have specific ranges of measurement of density. In calibration, the density of the reference metal may be situated in this range so that it is not possible to use mercury. Taking these circumstances into account, the methods of calibration of areometers require further improvement.

We believe that the most suitable and accurate methods are those based on the determination of the total volume of the areometer because these methods do not have the previously mentioned disadvantages. It is assumed that the radius of the bar along the entire length is constant, and the dependence of the volume of the immersed part of the areometer on the depth of immersion is presented in the form:

$$V = V_0 - \pi r^2 h \qquad (3.2.4)$$

where r is the radius of the bar, h is the height of the part of the bar of the areometer projecting above the liquid.

The total volume of the areometer may be determined with a considerably higher accuracy that the volume of the immersed section. This may be carried out using the pycnometric method of determination of the volume of solids [117].

Analysis of primary sources shows that gallium and tin are used often in calibration [145–147], assuming that the temperature dependence of these metals is determined with sufficient accuracy.

However, as shown in [143], the literature data on the density of pure metals, including gallium and tin, are not in good agreement. The scatter of the literature data is sometimes considerably higher than the total error of determination of density by different authors. The problem with the selection of reliable data in calibration is solved by the experimentator subjectively to a certain degree and this increases the error of the measurement not considered previously. Therefore, calibration using pure mercury should be regarded as most reliable.

It is important to take into account that one of the important tasks is the critical analysis of the existing experimental data on the density of metals and alloys. The main difficulty in this direction is the absence in the primary sources of the description of the process of calibration of measuring devices and the data used in calibration. In some cases, these data may be found in dissertations which are often difficult to access. We believe that without this critical analysis of the experimental data, any recommendation or recommended values [143] will be subjective.

In conclusion, attention will be given to the effect of the thermal expansion of the areometer on the measured density values. In most cases, calibration is carried out at room temperature and density is measured in a specific temperature range. Therefore, in accurate measurements of the temperature dependence of the density it is necessary to introduce a correction which takes into account the variation of the volume of the areometer.

Depending on the application and working conditions, the areometers, like pycnometers, are produced from glass, quartz, refractory metals and oxides and also other materials [148, 149]. We used areometers produce from molybdenum glass S52-1. According to the certificate data, the coefficient of expansion of the glass was $52.2 \cdot 10^{-7}$ deg^{-1}. Taking into account the volume change with increasing temperature, the equations (3.2.1) and (3.2.3) have the following form:

$$\rho = \frac{P_0}{g\left[\left(V_0 - \pi r^2 h\right)\left(1 + \beta \Delta T\right) - \pi r R \Delta H\right]} \qquad (3.2.5)$$

$$\rho = \frac{m_2 - \frac{r_2}{r_1} m_1}{\left(V_2 - \frac{r_2}{r_1} V_1\right)\left(1 + \beta \Delta T\right)} \qquad (3.2.6)$$

where the expression $(1 + \beta \Delta T)$ takes into account the thermal expansion of the immersed part of the body of the areometer, β is the volume coefficient of thermal expansion of glass. The equations may be used for the more accurate measurement of the temperature dependence of the density of metals and alloys, taking the thermal expansion of the areometer into account.

In conclusion, it should be mentioned that neglecting the correction for the thermal expansion of the areometer, results in appreciable errors in density measurements and, in particular, in its temperature coefficient. As an example, the data are presented on the density of liquid tin obtained taking into account and disregarding the thermal expansion of the areometer (g/cm³):

$$\begin{aligned} \rho_1 &= 7.002 - 7.0 \cdot 10^{-4} (T - 505) \cdot K \\ \rho_2 &= 7.026 - 5.9 \cdot 10^{-4} (T - 505) \cdot K \end{aligned} \qquad (3.2.7)$$

In the calculation of surface tension using these data, we obtain the following results:

$$\begin{aligned} \sigma_1 &= 544 - 0.070 (T - 505) \cdot K \quad (\text{mJ/m}^2) \\ \sigma_2 &= 544 - 0.070 (T - 505) \cdot K \quad (\text{mJ/m}^2) \end{aligned} \qquad (3.2.8)$$

As indicated by the equations (3.2.7) and (3.2.8), the errors in determination of the temperature coefficient of density, determined by the thermal expansion of the areometer, reach 16%, and the error in determination of the thermal coefficient of surface tension is 12%.

Thus, in accurate measurements the areometric method makes it possible to measure the density with a sufficiently high accuracy. However, to obtain the accurate data by this method, it is necessary to take into account strictly the effect of capillary forces and thermal expansion of the areometer.

3.3. Determination of the concentration dependence of the density of metallic melts

Because of the high rate of oxidation of liquid metals and alloys, the measurement of the physical–chemical properties of these materials should be carried out in vacuum or an inert medium. This results in considerable difficulties in the examination of the concentration dependence of the physical–chemical properties of binary and, in particular, multicomponent systems. The individual methods of preparation of the alloys and of measurement of the properties are associated with the time-consuming and intensive procedure of preliminary preparation of the device for individual measurements. In the mean, 3–5 days are required for the preparation of a device and measurement of the temperature dependence of density of a single alloy. This is accompanied by the extremely high consumption of expensive metals and materials. Therefore, it is interesting to develop methods and procedures which would make it possible to increase the productivity whilst maintaining the accuracy of measurements.

This task was formulated for the first time and solved by Pugachev and Lazarev in the examination of the surface tension and work function of the electron of ternary amalgams. The use of a microburette as a dosing device with preliminary calibration with respect to volume enabled the authors to prepare in the device, without disrupting the vacuum, tens of alloys of different concentration and measure their properties [127]. A shortcoming of these devices is that they do not make it possible to measure of the density of the melt.

The problem of the combined measurement of the surface tension, density and work function of the electron in the same thermal vacuum conditions was solved over a long period of time in our laboratory. These devices have been described in detail in [299]. Therefore, in this book, we describe only the devices which have been designed with direct participation of the author or used in the experimental investigations. Some of these devices, which make it possible to increase the productivity of experimental investigations, have been used in the experimental measurements of surface tension and density. Prior to describing these devices, attention will be given to several variants of increasing the productivity in the measurements of the concentration dependence of the density of melts.

3.3.1. Pycnometers for the measurement of the concentration dependence on the density of melts [150]

The first of these devices [150] consists (Fig. 3.1) of the cylinders 1–3, 6, 7, the microburette 9 and the pycnometer (cylinder 8 with the capillaries 4 and 5). The microburette and the pycnometer are calibrated in advance with respect to volume. The device may be used for the preparation of alloys of different compositions and for measuring their density, without opening the device or weighing. This is the main advantage of the device as a result of the fact that

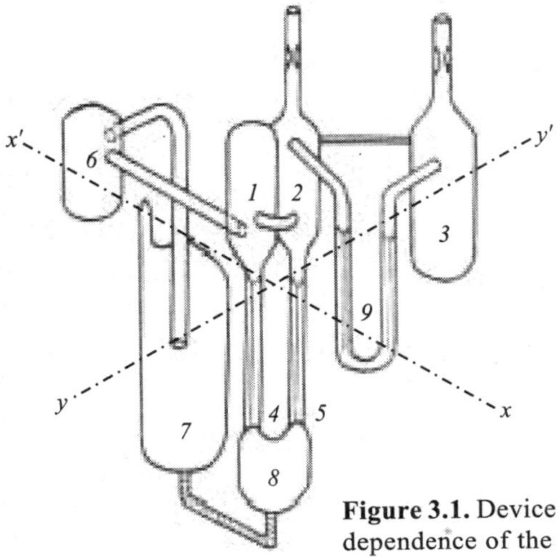

Figure 3.1. Device for the determination of the concentration dependence of the density of melts.

the mass of the alloy is not determined by the conventional weighing procedure, characteristic of the pycnometric method, and it is determined by the measurement of the volume of the components at the known density.

To determine the concentration dependence of the density of the liquid alloy (the components of the alloy are metals A and B, the density of these metals is known in advance), it is necessary to remelt the metal A in vacuum in a pycnometer, and the metal B or should be melted in the cylinder 3. Subsequently, the device is disconnected from the vacuum station and placed in a thermostat with inspection windows. If the density of the metal A is not known in advance, it is necessary to weigh the remelted amount of the metal and used the volume occupied by the metal in the pycnometers to determine its density by the conventional procedure. In the case of a multicomponent system, the metal should be replaced by the n–1 component alloy, to which the n-th metal B is added.

The amount of metal A (or of the multicomponent alloy) should be calculated to ensure that after remelting the capillaries 4 and 5 are partially filled. If the amount of the metal is larger, then, tilting the device around the axis y–y' in the anticlockwise direction through the angle of 90°, the excess of metal is transferred into 6. Further, the volume of the liquid metal is determined along the height of the liquid metal A in the capillaries of the pycnometer 4 and 5 and this is followed by the determination of the mass of liquid metal.

To prepare the first alloy, it is necessary to pour the part of the metal B from the cylinder 1 into the microburette 9. For this purpose, the device is inclined around the axis x–x' in the anticlockwise direction through the angle of 90°. In this case, it should be taken into account that the volume of the metal in the microburette should not exceed the part of the capillaries of the pycnometers free from the metal. If the amount of the metal in the microburette is too large, part of the metal is transferred back into the cylinder 1. After filling the microburette with the required amount of the metal B, the volume of the metal is determined and transferred into the microburette. In this case, the device is rotated around the axis x–x' in the clockwise direction. After mixing, carried out by multiple rotation of the device around the axis x–x' in the clockwise direction, and returning to the initial position, the alloy is ready for measurements. Since the mass of the alloy is equal to the sum of the masses of the components, to determine the density of the alloy, it is sufficient to measure the volume occupied by the alloy in the pycnometer.

The same procedure is used for the preparation of the alloys of other concentrations. For this purpose, using the method described previously, part of the alloy situated in the pycnometers, with the concentration and density determined in the previous measurements, is transferred into the cylinder 7.

The volume and, consequently, the mass of the part of the alloy transferred into the cylinder 7, are determined by the measurements of the density of the levels of the meniscus in the capillaries 4 and 5 prior to and after pouring. Subsequently, using the microburette, the required amount of metal B is added to the pycnometers and, determining its concentration, density measurements are started. The measurement of the volume, occupied by the metal in the pycnometers, makes it possible to determine its density.

The same procedure is used for the preparation of other alloys.

The intermediate cylinder 6 makes it possible to pour the metal or alloy back into the pycnometer, if a large amount of metal is poured from the pycnometers, *i.e.*, if the level of the remaining metal is not situated on the flat calibrated section of the capillaries.

The measuring section of the device is the two-capillary pycnometer [151] ensuring high accuracy of the results. One charge is sufficient for the examination of density with respect to the entire concentration range. This makes it possible to reduce greatly the consumption of metals and experiment time and ensure identical conditions of the measurement of the density of different alloys.

Shortcomings of the device include the fact that the area of every consecutive point includes the sum of the errors of all previous points.

3.2.2. Areometric method of determination of the concentration dependence of density

The concentration dependence of the density of a multicomponent system can be determined using an improved areometer. This includes the possibility of preparing an alloy without opening the device and disrupting the thermal vacuum conditions. One of these devices is described below.

The operation of the device (Fig. 3.2) is based on the following procedure. After preliminary preparation of the device for measurements, the investigated metals A and B are remelted in the tubes 1 and 2. This is followed by the measurement of the density of pure metals, if these values are not available. The density of metal A is measured by areometers 3 and 4. The measurement of the density of the second metal A is taken by the second pair of the areometers.

After determining the temperature dependence of the density of pure metals, the procedure of the preparation of the first metal is initiated. For this purpose, part of the metal B is added to the metal A. The microburette 8, after preliminary calibration with respect to volume, is used for the determination of the volume and, consequently, the mass of the added part of the metal B.

The resultant melt is intensively mixed by the appropriate rotations of the device. The density of the result and homogeneous

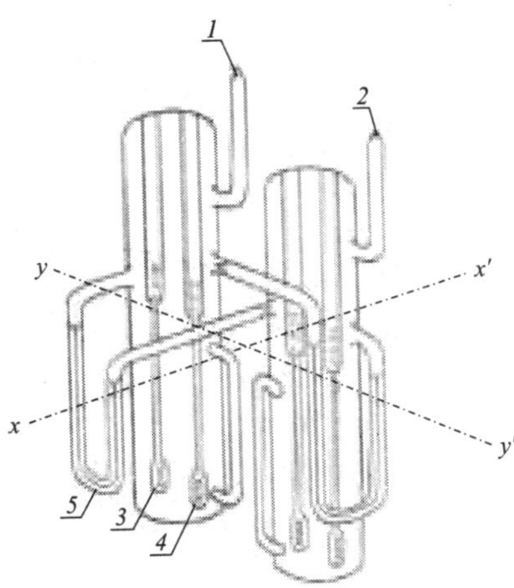

Figure 3.2. The areometric method of determination of the density of melts.

melt is measured, as in the case of the pure metal A, using the areometers 3 and 4.

The second alloy is prepared in the areometric chamber 2. In this case, a certain amount of the first alloy, after measuring its density, is added to the metal B using the microburette 5. For this purpose, the devices are positioned around the axis y–y' under the angle of approximately 90° in the clockwise direction so that the certain amount of metal is transferred into the microburette 5 and, determining its mass and volume by rotating the device around the axis y–y' in the anticlockwise direction, is added to metal B. After thorough mixing, the density of the resultant alloy is determined using the second pair of the areometers.

Certain difficulties are encountered in the determination with a single areometer of the density of the melt, whose components greatly differ from each other in density. They are determined by the need for expansion of the ranges of measurement of the density of areometers. For this purpose, it is necessary to increase either the length of the areometer or the radius of the bar of the areometer. This reduces the sensitivity and, consequently, accuracy of the areometers.

This device is free from the above-mentioned shortcoming because two areometers are used for measuring the density. If the values of the density of the pure components greatly differ from each other, one pair of the areometers is used for measuring higher values of the density, and the other pair for lower values.

It should be mentioned that these devices may be used for measuring the angle of wetting by metals and glass melts. This is carried out using capillary discharge tubes, and the wetting angle may be calculated from the level of metal in these tubes in comparison with the level in the areometric cylinders.

Thus, the previously described devices may be used efficient if examination of the concentration and temperature dependence of the density of two-component metallic melts. These devices may also be useful in examining n-component solutions. For this purpose, one of the areometric cylinders or a pycnometer must be filled with $(n-1)$-component solution, and the second chamber of the areometer or the cylinder 3 of the pycnometric device should be filled with the added metal.

The main advantage of these devices is that they make it possible to prepare melts of different concentrations and measure the density of these melts at different temperatures, without opening the device. A single charge is sufficient for examining the concentration dependence of the density of a binary system (or a section of a multicomponent system) in the entire concentration range. Consequently, it is possible to reduce greatly the consumption of metals, experiment time, and also ensure identical measurement conditions. On the other hand, as a result of the development of these devices, it is now possible to design more complicated combined devices for the combined measurement of surface tension, density and electron work function. The design features of these devices will be examined later.

3.4. Combined measurement of the physical and chemical properties of liquid metals and their alloys

In the previous sections, attention was given to the devices and methods for measuring surface tension and density. In density measurements, it is possible to prepare in the device (without opening it) alloys of different compositions thus greatly increasing the productivity of equipment. However, in these devices the surface tension and density are measured separately. The experimentator faces the following difficulties:

– it is almost impossible to prepare alloys of the completely identical concentration in the pycnometers and in the device for the measurement of surface tension;

– thermal vacuum conditions of measuring the surface tension and density differ;

– the consumption of expensive metals and experiment time greatly increase.

The next stage in the development of high-productivity experimental systems for the measurement of surface tension and density is the design of devices for the combined measurement of these parameters. Several variants of these devices are described below.

3.4.1. A device for the combined determination of surface tension and density

Figure 3.3 shows the device combining the 'gravitation' procedure proposed by Pugachevich, and a two-capillary pycnometer. The concept of this combination was proposed for the first time in our laboratory [119]. We have used this device in practice in the examination of the ternary systems using the method of

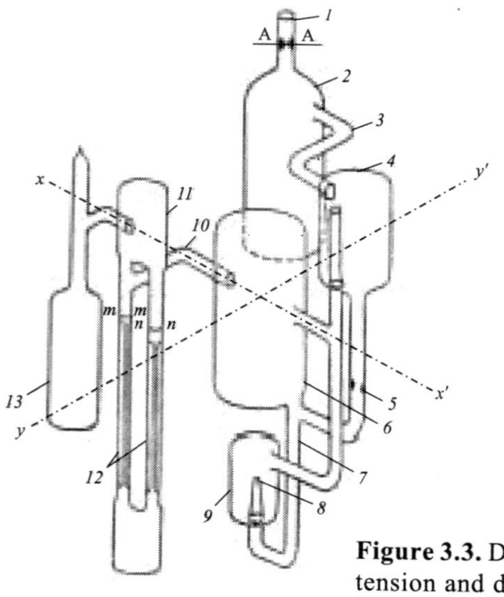

Figure 3.3. Device for combined determination of surface tension and density.

mathematical experiment design. The devices used for measuring the surface tension and density of the indium–gallium, tin–thallium binary systems and indium–tin–thallium, indium–lead–thallium and tin–lead–thallium systems.

The alloy to be investigated is transported into the device (Fig. 3.3) under vacuum through the pipe 1. Subsequently, the device is separated along the line A–A from the vacuum system and placed in an air thermostat with the examination windows. After reaching the required temperature in the thermostat, rotating the device together with the thermostat around the axis y–y' in the clockwise direction through the angle of 90°, the melt is transferred from the container 2 through the pipe 3 into the cylinder 4. From the latter of the melt gradually flows, through the narrow section 5, into a pressure gauge cylinder and through the pipe 7 into the capillary with the known radius 8. A droplet of the melt forms at the end of the capillary.

The height of the metal column in the pressure cylinder in relation to the end of the capillary, required for the separation of the first droplet, is the measure of excess pressure in the droplet and is recorded using a cathetometre through the inspection windows in the thermostat.

In the repeated tilting of the device in the anticlockwise direction, the metal is returned to the cylinder 4 and the entire process is repeated.

To measure the density, part of the melt is transferred from the cylinder 6 by rotating the device in the anticlockwise direction around the axis y–y' through the pipe 10 into the two-capillary pycnometer 11 with the calibrated volume.

Measuring the height of the melt in the capillaries 12 in relation to the marks m–m and n–n using the cathetometre, it is possible to determine the volume of the melt in the pycnometers at different temperatures.

After completing the measurements, the melt is transferred from the pycnometers into the cylinder 13 by rotating the device around the axis y–y' in the clockwise direction through the angle of 360°. Part of the melt, remaining in the cylinders 9, 6, 4, is transferred into the container 2. After solidification and cooling the melt to room temperature, the alloys from the cylinder 13 and the container 2 are extracted and the density of the alloys is determined from the mass of the alloy, extracted from the cylinder 13. The cylinder 13 and container 2 are reconditioned for subsequent measurements. After rinsing, drying and thermal vacuum treatment, the device is ready for further measurements.

3.4.2. A combined device for detailed examination of the physical-chemical properties of multicomponent melts

The device described below is designed for the combined measurement of the surface tension and density of metallic nose. As in the previously described device, the density is determined by the pycnometric method. However, a distinguishing feature of the devices is that it is suitable for preparing alloys of different concentrations, without the need for opening the device and disrupting thermal vacuum conditions in it. This is possible because of the combination of the 'gravitation' device with the pycnometer, described in section 3.3 (Fig. 3.4).

The operating principle of the device, shown schematically in Fig. 3.4, may be described as follows. After the thermal vacuum treatment of the device, a specific amount of metal A is remelted in the cylinders 1 and 2, and metal B in the pycnometer 14. Subsequently, the device is separated by unbrazing from the vacuum station and transferred into the air thermostat.

Initially, measurements are taken of the surface tension of metal A whose temperature dependence is not known in advance. For this purpose, the device is rotated in the anticlockwise direction through the angle of 90° around the x–x' axis in such a manner that the metal from 1 is transferred into the cylinder 4. When the device

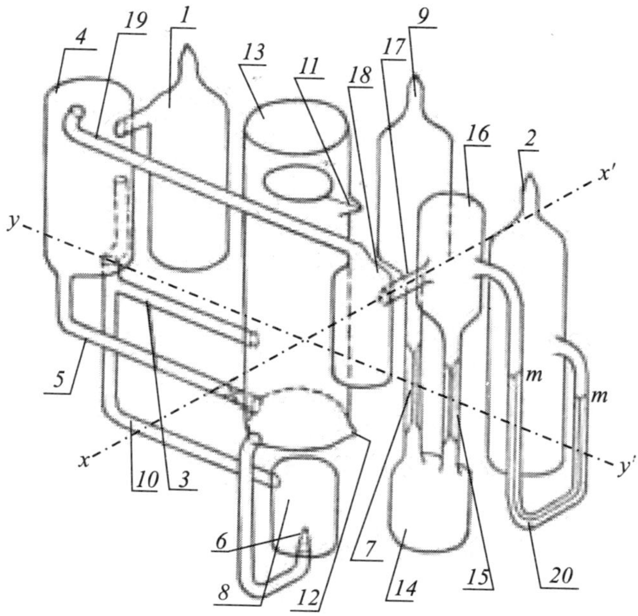

Figure 3.4. Device for detailed investigations of the physical–chemical properties.

is returned to the initial position, the metal A flows through the tube 5 into the pressure cylinder and the capillary with the known radius. Surface tension is determined by the same procedure as described in section 3.3. To repeat the measurements, it is sufficient to rotate the device in the anticlockwise direction to ensure that the metal passes through the tube 10 and 3 into the container 4. Returning the device to the initial position, measurements are repeated.

Since the metal B is remelted in the pycnometer, knowing its mass and determining its volume, it is quite easy to measure the temperature dependence of the density of this metal.

Thus, the device is used for the measurement of the surface tension metal A and the density of metal B. This was followed by the preparation of the first alloy. For this purpose, calibrated pipes of the pycnometer are used as the microburette–dosing device. Measuring the level of metal B in the capillary pipes, the device is rotated around the axis y–y' in the anticlockwise direction to ensure that the metal from the pycnometer 14 is transferred into the cylinder 16 through the capillary pipe 15 and subsequently into 18. Returning the device to the initial position, the level of the metal remaining in the pycnometer is measured. It is quite easy to determine the volume and mass of the metal from the difference of the levels prior to and after transfer. After supplying the required amount of metal into the intermediate cylinder 18, it is added to metal A. For this purpose, the device is rotated around the axis x–x' in the anticlockwise direction, and the metal from 18 flows into the cylinder 4.

The melt of the first concentration, prepared by this procedure, is investigated using the same procedure as the pure metal A. The melts for the measurement of density are prepared in the pycnometer. This is carried out using the volume-calibrated microburette 20 used for supplying the required amount of metal A from the cylinder 2 into the pycnometer 14.

To fill the microburette with the metal A, the device is rotated around the axis x–x' in the clockwise direction to such an extent that a certain amount of metal A flows into the microburette 20. Returning the device to the initial position, we determine the level of the metal in the microburette and, consequently, its volume. If the amount of metal in the microburette is large, part of the metal may be transferred back into the cylinder 2. Thus, the required amount of the metal should be transferred into the microburette and after determining its volume, it should be added to the pycnometer. For this purpose, it is sufficient to rotate the device around the axis x–x' in the anticlockwise direction.

The mass of the alloy in the pycnometer may be easily determined because the amount of the poured metal B and of the added metal A is known.

Continuing the investigations, the device may be used for the preparation of alloys of different concentrations and for examining their density and surface tension.

The device may also be used for measuring the work function of the electron using the procedure described in [61]. For this application, the device is fitted with the flat quartz window 13 and electrodes 11 and 12.

As shown by the description of the device, the melts for the measurement of surface tension and density are prepared in different parts of the device and their compositions differ. This is one of the shortcomings of the device. The values of the densities, required for calculating the surface tension, may be determined from the dependence of density on the composition which is determined with sufficient accuracy in the device. The main shortcomings, restricting the application of the device in practice, are the large size of the device and the buildup of errors. As mentioned previously, the error of measurement of the density of the given alloy includes the sum of the errors of measurement of all previous alloys.

3.5. Combined device for the determination of the concentration dependence of the surface tension and density of multicomponent melts [154]

In this section, we describe a device designed for the examination of the surface tension and density of multicomponent systems. We use the system for the measurement of the surface tension and density of all pure metals and a majority of the binary and multicomponent systems. Therefore, the design features of the device can be examined in greater detail. In the device, surface tension is measured by the method of maximum pressure in the droplet, by the 'gravitation' method, and the density is measured with an areometer taking into account the capillary forces and thermal expansion [154].

The diagram of the device produced from glass or quartz is shown in Fig. 3.5. After long-term (16–20 hours) pumping of the device, with simultaneous heating to 773–793 K, the required

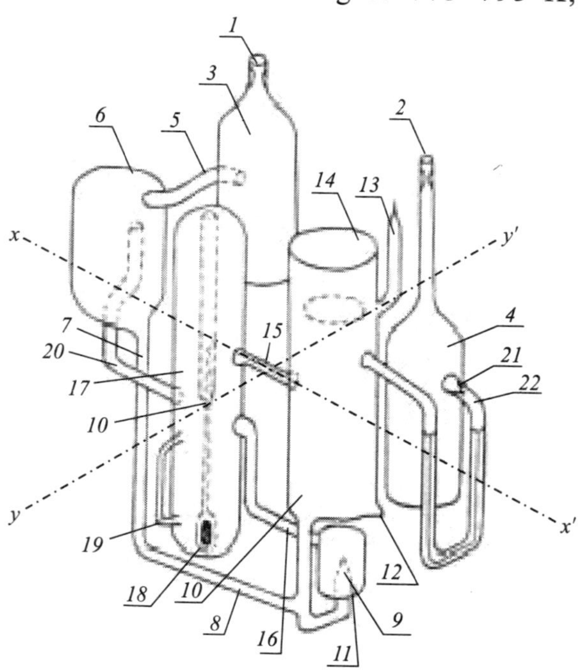

Figure 3.5. Combined device for determination of the surface tension and density of multicomponent systems.

amounts of the liquid metals and alloys are transferred into the cylinders 3 and 4 through the pipes 1 and 2. The device is then separated from the vacuum station and placed on a metallic frame in an air thermostat with inspection windows.

Initially, the surface tension and density of the melt transferred into the cylinder 3 are measured. For this purpose, the required temperature is established in the thermostat, the system is held at at this temperature for two hours and, rotating the device around the axis y–y' in the anticlockwise directions through the angle of 90°, the melt is transferred from the cylinder 3 through the pipe 5 into the cylinder 6. After returning the device to the initial position, the melt slowly fills the capillary 9 and the pressure cylinder 10 through the pipe 7 and the constricted section 8. A droplet forms at the outlet of the capillary with the known radius. The height of the melt column required for separation of the droplet, is measured with a cathetometer. It is equal to the difference between the maximum level of the melt in the cylinder 10 and the outlet of the capillary 9. Knowing, in addition to this, the radius of the capillary and the density of the melt, it is quite easy to calculate surface tension. After separation of the first droplet, the metal rapidly fills the cylinder 11 and partially cylinder 10.

To measure the density of the melt, part of the melt is transferred from the cylinders 10 and 11 through the pipes 15 and 16 into the areometric cylinder 17. For this purpose, the device is rotated around the axis y–y' in the anticlockwise directions through the angle of 90° and subsequently returned to the initial position. The density of the melt is determined using areometer 18. The capillary outlet 19 is used for taking into account the effect of capillary forces. The capillary outlet is used instead of the second areometer due to the fact that this variant proved to be more suitable for the same accuracy of measurement. At the same time, if necessary, the capillary outlet may be replaced by a second areometer.

For repeated measurement of surface tension of the melt, the device is rotated around the axis y–y' in the anticlockwise direction to such an extent that the melt is transferred from the cylinders 10, 11 and 17 into the cylinder 6 to the pipe 20. Whether returning the device to the vertical position, the melt again fills the capillary 9 and the pressure cylinder 10. Repeating the procedure, the measurements may be repeated as many times as necessary.

After completing examination of the surface tension and density of the metal or alloy transferred into the cylinder 3, the alloy with the next concentration is prepared. For this purpose, the device is rotated around the axis y–y' to such an extent that the required amount of the second metal from the cylinder 4 is transferred into the microburette 23 through the capillary 21. Returning the device to the initial position, we determine the volume and, from the known value of density, the mass of added metal. Subsequently, by the appropriate rotation of the device, the metal is transferred from the microburette into the cylinder 10. The resultant alloy with the new concentration is thoroughly mixed, transferred into the cylinder 6 and back to 10. Measurements are then started. The same procedure is used for the preparation of alloys of new concentrations and for the measurement of their properties.

The device may be used for measuring the work function of the electron using the electrodes 12, 13, and the quartz window 14.

After completing the measurement, the melt is transferred from all parts of the device into the cylinder 6 and subsequently 3 and the device should then be rotated through the angle of 360° around the axis y–y' in the anticlockwise direction. After cooling the system, the cylinder together with the metal are cut-off from the device and a new device is fixed to it. The device is then ready for new measurements.

Depending on whether pure metals or alloys are remelted in the cylinders 3 and 4, the device is used to examining binary or multicomponent melts.

As indicated by the description, the device may be used for the preparation (without opening and disruption of the thermal vacuum conditions) of alloys of different concentrations and for measuring the surface tension, density and work function of the electron of these melts. The application of the device reduces by a factor of 10 the consumption of expensive metals and increases the productivity of experiments by the same amount. The labour content of the experiments is also considerably lower because the experimentator is freed from the procedures of rinsing, drying and long-term vacuum thermal treatment of the device prior to the measurement of the properties of each alloy. In addition to this, the measurement of the parameters in the identical conditions results in efficient comparison and determination of a probable correlation between them.

As mentioned previously, in the examination of the properties of the multicomponent systems in a wide concentration range, it is very important to increase the productivity of the device whilst retaining the given accuracy. It is also important that the conditions of measurement of the surface tension and density of alloys of different alloys are identical with respect to the thermal vacuum treatment of the device, the purity of the initial components and thermostatic control of the melt. Consequently, in examination of the properties of multicomponent systems, it is obviously preferred to use the combined devices. It should be mentioned that these devices can be used when one of the components of the melt has a high vapour tension at the measurement temperature. In this case, there may be uncontrolled exchange of matter through the vapour phase between the melt in the measuring section and the added metal. However, as shown later, in the case of metals investigated here, the effect of this factor is insignificant.

3.6. Experimental equipment for the methods of measurement of the physical–chemical properties of multicomponent metallic melts

The application of combined devices is characterised by a relatively long measurement time (3–5 days). During this period, the device, separated from a vacuum system, is placed in a thermostat whose temperature continuously changes from the liquidus temperature to 773 K. In these conditions, the vacuum in the device is impaired because of the generation of residual gases from the internal surface of the device and the metal. Therefore, in operation with the combined devices, special requirements are imposed on the process of preliminary thermal vacuum treatment.

Special experiments show that after conventional thermal vacuum treatment for 4–6 hours, the reduction of the efficiency of vacuum may already be noted in the second day after the moving the device from the vacuum station. This circumstance requires careful preliminary thermal vacuum treatment of the device and investigated metals. The equipment for thermal vacuum treatment and loading the device should minimise the amount of secondary substances in the measurement cell.

For experiments, we assembled experimental equipment, with the main sections being: a vacuum station, producing a rarefaction of the order of $5\cdot10^{-5}$ Pa in the device in continuous cooling with nitrogen traps; an air thermostat with inspection windows in which the temperature can be regulated from room temperature to 823 K, and the maximum temperature gradient in the region in which the device is located does not exceed 0.05 K/cm; a furnace for thermal vacuum treatment of the device with the temperature controlled in the range from room temperature to 1000 K; a bunker for heat treatment and filtration of metal; KM-8 cathetometre; electric power supply for experimental equipment, a glass combined device, described previously. The vacuum metres VIT-1A and VIT-14 with the sensors LT-2 and LM-2 were used for measuring the vacuum.

Glass is capable of adsorbing various gases at room temperature. In heating the device, the gases are desorbed and reduce vacuum. The desorbed gases, if they are not removed in advance, may greatly influence the experimental results [130]. Therefore, special attention will be given to the problem of loading the system with the investigated metals and preliminary thermal vacuum treatment of the device.

The diagram of the experimental equipment is shown in Fig. 3.6.

The device and all glass sections of the vacuum system were rinsed with a hot chromium mixture and distilled water and subsequently dried [155]. The rinsed and dried device was fixed on a metallic frame and placed in the strictly vertical position in the thermostat. Subsequently, the metal frame together with the device were transferred into the furnace for heat treatment and fixed securely on the frame.

The metals, after mechanical removal of oxides and weighing on an analytical balance, were placed in the bunker 4 for thermal vacuum treatment.

The following metals were used in the investigations: grade 0VCh-000 tin, IN-00 indium, S-0000 lead, Tl-00 thallium, VI-000 bismuth, Gl-000 gallium.

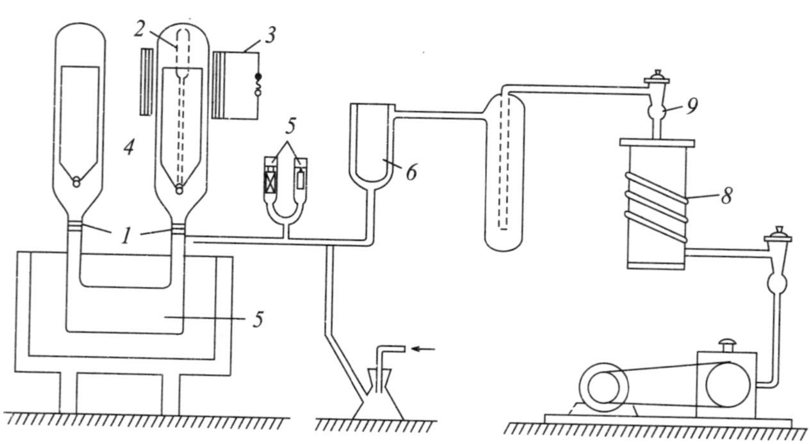

Figure 3.6. Diagram of experimental equipment.

The quartz bunkers contain support rings, funnels with orifices of approximately 0.5 mm in diameter, and magnetic valves 2 closing the capillaries through which the metal flows into the device. In the melting of the metals, the magnetic gate is opened using the electromagnet 3. The bunkers, the magnetic gates and the funnels are made from quartz so that the metals may be degassed at high temperatures. On the other hand, the application of the magnetic gates allows the preparation (if necessary) of homogeneous alloys in bunkers followed by remelting them in the device. This is important in the examination of multicomponent systems because it eliminates the possibility of disrupting the concentration of the binary alloy remelted in the device, and therefore ensuring the constancy of the given ratio of the component. Through the quartz-molybdenum transitions I, the bunkers are connected with the device and the entire system to the vacuum station.

The thermal vacuum treatment of the device is carried out using the following procedure. Initially, pumping is carried out using a rough a vacuum pump and the entire system is cleaned in plasma [156] generated by a high frequency discharge. This is followed by activation of an oil vapour pump. When the vacuum of 10^{-5} Pa is released in the system, the temperature is increased to 773 K. With the increase of temperature, as a result of the generation of residual gases from the surface of the glass, and the vacuum in the system is reduced and then starts to increase after a certain period of time. When the maximum vacuum is reached, liquid nitrogen is added to the trap and this increases the vacuum to 10^{-5} Pa. This sequence increases the efficiency of the effect of nitrogen traps.

In order to degas the metals, the temperature in the cylindrical furnaces increases to 950 K. A further increase of temperature is not desirable because not only light gases but also oxygen penetrate at high temperatures through the quartz used for the production of the bunkers [157].

Temperature is measured with a chromel–alumel thermocouple during degassing of the metals and with a mercury thermometer in the measurement of the physical–chemical properties.

After pumping for 12–14 hours, the cylindrical furnaces were cooled to a temperature slightly higher than the melting point of the metals, and the metals were remelted in the device. Subsequently, magnetic gates were opened using an electromagnet. The metals, filtered off in funnels, were transferred into the measuring device. In order to protect the device against the penetration of oxides, a small amount of the metal was not remelted.

The oxides, being surface-active admixtures, greatly change the values of surface tension even at a small concentration. Therefore, special attention must be paid to the purification of metals prior to charging. Here, it should be mentioned that the amount of the oxides, remelted together with the metal, also depends on the remelting temperature. It has been found that with the increase of temperature, the oxides, adsorbed on the surface, are desorbed and this reduces the quality of filtration by capillary constrictions. With a decrease of temperature the oxides concentrate mainly on the surface layer and are effectively retained during passage of the metals through the narrow section of the capillary. Therefore, the remelting of metals is carried out at temperatures slightly higher than the liquidus point.

To reduce the pumped volume in the area from which the gases are released, the quartz tube together with the bunkers are disconnected by unbrazing after remelting the metals. The device with the metals is heated and continuously evacuated for 3–4 hours. The vacuum increases to $5 \cdot 10^{-6}$ Pa.

The last and, in fact, the most important moment, is the separation of the device from the vacuum system by unbrazing. This is carried out using a special constriction (a constricted thick wall glass tube). The unbrazing should be carried out at constant pumping whilst retaining the maximum vacuum in the device. Prior to unbrazing with the flame of a manual torch, the constricted section is heated several times almost up to the softening temperature of the glass. This is carried out to remove the adsorbed gases from the glass. The end of degassing was determined on the basis of the establishment of vacuum in the system. Subsequently, the device was disconnected from the vacuum system. The device, secured to the metallic frame and disconnected from the vacuum system, is transferred into an air thermostat for measurements.

3.7. Method of processing measurement results

3.7.1. The law of distribution of the error of measurement of surface tension

For the accurate evaluation of the measurement errors, it is necessary to have accurate information on the distribution function of errors [158, 159]. Usually, this is carried out on the basis of the

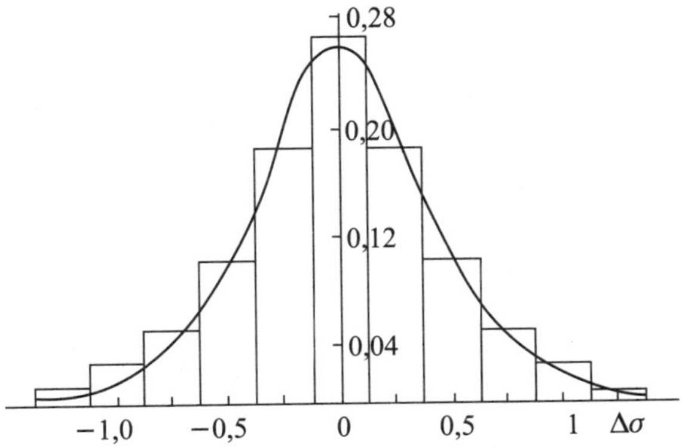

Figure 3.7. The distribution function of errors of measurement of surface tension.

normal distribution of the sample. However, no strict confirmation is available for these assumptions. Theoretical assumptions, based on the central limiting theorem of the probability theory, are not always valid. In particular, the theorem is not fulfilled when one of the random errors is dominant. Therefore, the experimental verification of the hypothesis on the form of the distribution function of the random error is of specific interest. As far as we know, the problem of distribution of the error of measurements of surface tension by the method of the maximum pressure in the droplet is examined in this book for the first time.

For the construction of the accurate function of distribution it is necessary to use a large number of experimental data. Therefore, we have used our own data on the surface tension of low melting alloys.

Since every measurement is repeated several times, we have samples of 8–10 measurements with different values of the mathematical expectation. The type of the distribution function does not depend on the value of mathematical expectation. Consequently, the individual samples may be converted to the zero value of mathematical expectation. The distribution function retains its form. Consequently, statistical material can be selected from a large number of observations. To plot the histogram shown in Fig. 3.7, we have used approximately 8000 measurements.

In addition to the experimental histogram, the figure shows the theoretical curve of the distribution function of the errors, which indicates that the theoretical curve describes the experimental data quite satisfactorily. It also indicated by the results of our quantitative calculations using the Pearson K^2-criterion [161]. The resultant histogram shows that the surface tension in the gravitational device is determined with high accuracy (dispersion is equal to 0.32 mJ/m^2).

Thus, the random errors of measurement of the surface tension by the method of maximum pressure in the drop are distributed in accordance with the normal law. However, the dispersion is caused by both the random and systematic components:

$$\sigma^2 = \sigma_{syst}^2 + \sigma_{rand}^2 \tag{3.7.1}$$

Increasing the number of observations, the random component may be reduced to such an extent that it can be ignored. Therefore, the main role in the evaluation of the measurement error is played by the systematic error. This error is influenced by the following factors: the error of calibration of the internal radii of the capillaries, the presence of ellipticity on the internal cross-sections of the capillaries, errors in reading the height of the meniscus, the diameter of the vessel of the pressure cylinder, wetting of the glass by the melt, optical distortion because of the irregularities on the external walls of the device, determination of the height of the meniscus before the period required for reaching the maximum pressure, vertical position of the capillary in the measuring devices. Therefore, below we present the estimate of the systematic error of the dispersions which determines the error of measurement of surface tension and density.

3.7.2. Error of determination of the composition of the melt prepared using a dosing apparatus

The alloy is prepared in the combined device using a pre-calibrated microburette. We determined the error of determination of the mass of the added metal and, consequently, the composition of the investigated melt. The mass of the metal, added to the microburette, is determined from the density and volume occupied by the metal in the microburette:

$$m_i = \rho_i A h \tag{3.7.2}$$

where ρ_i is the density of the metal; h is the height of the level of the metal in the microburette; A is the cross-sectional area of the capillary from which the microburette is produced. According to the statistical law of buildup of errors, the dispersion is equal to:

$$\frac{S_{m_i}^2}{m_i^2} = \frac{S_\rho^2}{\rho^2} + \frac{S_A^2}{A^2} + \frac{S_h^2}{h^2} \tag{3.7.3}$$

The volume of the microburette up to the given mark is determined by preliminary calibration. Mercury was used for calibration:

$$V = hA = \frac{m_{\text{Hg}}}{\rho_{\text{Hg}}} \quad \text{consequently} \quad A = \frac{m_{\text{Hg}}}{\rho_{\text{Hg}} h} \tag{3.7.4}$$

Consequently

$$\frac{S_A^2}{A^2} = \frac{S_{m_{\text{Hg}}}^2}{m_{\text{Hg}}^2} + \frac{S_{\rho_{\text{Hg}}}^2}{\rho_{\text{Hg}}^2} + \frac{S_h^2}{h^2} \tag{3.7.5}$$

Substituting (3.7.5) into (3.7.3) we obtain:

$$\frac{S_{m_1}^2}{m_1^2} = \frac{S_\rho^2}{\rho^2} + \frac{S_{m_{Hg}}^2}{m_{Hg}^2} + \frac{S_{\rho_{Hg}}^2}{\rho_{Hg}^2} + 2\frac{S_h^2}{h^2} \tag{3.7.6}$$

This equation shows that the accuracy of dosing depends both on the accuracy of the daytime the density of the added metal and on the error of determination of the mass and the height of the metal: in the microburette.

Into the equation (3.1.16) we substitute the values for the quantities in the equation and carryout calculation. This gives

$$\frac{S_m}{m} = 0.002$$

Assuming that the mass of the smallest portion of the addition is 1 g, it is quite easy to find the relative area of determination of the concentration $\frac{S_c}{C}$ =0.0022. Consequently, dosing of the alloys using the microburette may be carried out with the accuracy of 0.022% for the alloy with the concentration of 1% and with the accuracy of 0.2% for the alloys with the equimolar composition (increase of the error with the increase of the content is determined by the buildup of the error with the addition of a new portion of metal). The resultant accuracy of determination of the composition is fully sufficient for solving the problems formulated in the study.

3.7.3. Error of the measurement of the density of liquid metals

The error of the measurement of density with the areometer is calculated using the equation (3.2.1). In deriving the equation, it was assumed that the bar of the areometer and the capillary discharge are strictly cylindrical. The form of the real capillaries and bars differs from cylindrical and mean values of their radii are used in practice. This results in an error in the measurement of the density which is difficult to take into account accurately. Therefore, its maximum value is evaluated assuming that the error of measurement of the radius is not the error of the measuring device but the ellipticity of the radius of the bar of the areometer and capillary discharge tubes. The ellipticity for the diameters of the bars and capillaries, used in our experiments, was $\Delta D = 0.002$ cm.

The errors, caused by the inaccurate determination of β, g, T, do not provide any significant contribution to the error of density measurements.

The quantities h and Δh are measured with a cathetometer, with the error of $S_h = 0.0005$ cm.

Calculations of the error of measurement of density show that $\frac{S_\rho}{\rho} \cdot 100\% = 0.2\%$. This accuracy of determination of density is fully sufficient for calculating surface tension. Assuming that $\rho = 10$ g/cm^3, gives $S_\rho = \pm 0.02$ g/cm^3 and, consequently, for the confidence range 0.04 g/cm^3 the reliability of the density measurements is 95%.

The density of some of the melts was measured with a pycnometer. The values of the errors formed in this case will be estimated. Mercury was used for the calibration of the pycnometer in a vacuum at room temperature. The dependence of the volume on the height of the level of the mercury in the capillaries was approximated by a linear function using the method of least squares.

Calculation showed that the error of determination of the volume on the basis of the density and mass of mercury is $\frac{\Delta V}{V} = 10^{-4}$ cm^3.

An important source of errors in the pycnometric determination of the density is, as in the case of the areometer, the deviation of the true form of the capillaries from cylindrical. The accurate evaluation of the error is difficult in this case. However, the errors, introduced by these deviations and also by other factors not taken into account, may be estimated approximately by the mean quadratic deviation of the individual measurement, obtained in processing of the calibration data by the method of least squares. For the pycnometers used in our investigations $S_V = 0.01$ cm^3. This error is considerably greater than both the previously mentioned errors and also those formed as a result of the imperfection of the marks on the pycnometers, deviations from the vertical position, the error of the cathetometer, *etc*. This error is also used for defining the pycnometer $S_\rho = 0.01$ g/cm^3.

Thus, the error of determination of the density with a pycnometer is slightly higher than with the areometer. However, as shown previously, this shortcoming of the areometric method of density measurement is compensated by other advantages of this device in comparison with the pycnometer and the selection of a specific method of measuring density depends on the task in hand.

3.7.4. Error of measuring surface tension

The surface tension is calculated using equation (3.1.2) in which the third term is not taken into account because it is too small. An error, whose magnitude depends on the radius of the capillary, forms in this case. Since at $r \ll 0.3a$, where a is the capillary constant, the error is smaller than 0.3% [116]. The total error of measurement of the surface tension is determined by the expression:

$$\frac{S_\sigma}{\sigma} = \sqrt{(\delta\sigma)^2 + \left(\frac{S_r}{r}\right)^2 + \left(\frac{S_\rho}{\rho}\right)^2 + \left(\frac{S_h}{h}\right)^2} \qquad (3.7.7)$$

The maximum height of the metal column is measured many times. In our measurements, the coefficient of variation of this quantity at a fiducial probability of 0.95 was 0.1%.

The real capillary has an ellipticity whose value can be easily determined from the formula:

$$\Delta = \frac{r_{max} - r_{min}}{r_{cp}}$$

This introduces a systematic error into the measurement of surface tension. It is difficult to take this error strictly into consideration, because of the absence of a reference value for surface tension. The maximum possible value of this error is determined if it is assumed that the error of measurement of the radius of the capillary is equal to its ellipticity. The ellipticity of the capillaries, used in our experiments, did not exceed 0.6%. Therefore, it is assumed that $S_r = 0.006$.

The values of the quantities included in the calculation formula and the calculations of the maximum total error of measurement of surface tension show that the coefficient of variation of surface tension at a fiducial probability of 0.95 is 0.8. Assuming that the surface tension is on average equal to 500 mJ/m^2, for the RMS deviation we obtain $S_\sigma = \pm 4$ mJ/m^2.

We have calculated the maximum possible error of measurement of surface tension. It should be taken into account and the scatter of the experimental data for the given device is characterised by two last terms in (3.7.7) and this equals 0.2% or 1 mJ/m^2. In examination of the temperature or concentration dependence, this error is of special interest.

Analysis of the literature data on the surface tension of the metals shows that the scatter of the experimental data for the same metal, obtained by different authors, is considerably higher than even the total error of measurements. Evidently, these contradictions are caused by the fact that the experimentators do not always indicate the maximum possible error. By strict consideration of the maximum possible error it would be possible to solve a number of contradictions.

4

Surface properties and molar volumes of pure metals and binary systems

Regardless of the large number of investigations, concerned with the examination of the surface tension of binary systems, the reliability of data for some systems causes doubts. In a number of cases, more detailed repeated measurements indicate the absence of the previously detected extrema of the isotherms of surface tension [165–168].

To obtain the functional dependence of the surface tension and density on the composition of the multicomponent system, it is not sufficient to have the reliable data only on the multicomponent melts. A similar important role is played by the data on pure components and lateral binary systems, forming the multicomponent system. This problem is especially acute in the examination of multicomponent systems by the methods of mathematical experiment design. In this case, it is very important to minimise the scatter of the experimental points.

If pure metals, binary and multicomponent alloys are examined by the same device, the scatter of the experimental points is determined by the random error which may be minimised by increasing the number of measurements. The use, in this case, of literature data on the pure metals and binary systems may result in disagreement in the experimental data, obtained by different authors. Even if the same experimentator uses different devices in examining the same system, the scatter of the experimental data may increase. This may be due to the fact that the systematic errors obtained by different devices have the opposite signs and it is therefore necessary to add up the random error and, consequently, increase the scatter of experimental data.

Therefore, we have measured repeatedly the temperature dependence of surface tension and density of pure metals which are components of the multicomponent systems investigated in our work, and of lateral binary systems, forming the investigated multicomponent systems.

In this chapter, the experimental data are presented on the surface tension (σ) and density (ρ) of pure metals: indium, tin, gallium, thallium, lead and bismuth. We also discuss the results of examination of the surface properties and molar volumes of binary systems: indium–tin [211], thallium–lead [311], thallium–bismuth [211], thallium–indium [168], thallium–tin [310] and gallium–bismuth [96].

The following metals were used in the experiments: grade OVCh-000 tin; In-00 indium, C-0000 lead; Tl-000 thallium; Bi-000 bismuth; Gl-000 gallium. Measurements were taken in the temperature range from the liquidus temperature to 773 K in the combined device, described in section 3.5.

Using the method, described previously (chapters 1 and 2), and the experimental data on σ and ρ, we calculate the main thermodynamic parameters of the surface layer of the experimentally examined metals and alloys: the effective thickness of the surface layer, the molar area, the adsorption of components and the composition of the surface layer. The results are used to attempt the qualitative explanation of the special features of the isotherms and polytherms of the surface tension of the investigated binary alloys.

4.1. Temperature dependence of the parameters of the surface layer of low-melting alloys

The temperature dependence of the surface tension of even pure metals has been studied insufficiently. The accuracy of determination of the temperature coefficient of surface tension $\left(\frac{d\sigma}{dT}\right)$ is almost an order of magnitude lower than the accuracy of measurement of surface tension. This is caused by the following reasons: low values of $\left(\frac{d\sigma}{dT}\right)$, the strong effect of surface-active impurities on the temperature dependence of surface tension, and a narrow temperature range of measurements. Therefore, in accordance with the experimental procedure, described in chapter 3, we have increased the requirements on the experimental conditions (the purity of initial metals, the pressure of residual gases, preliminary preparation of the device). Measurements were taken in the combined device. Surface tension was determined from the maximum pressure in the droplet and the density was determined using an improved areometer taking into account the effect of capillary forces and thermal expansion. In the investigated temperature range, the dependence of the density of metals on temperature is linear with a negative temperature coefficient. Processing of the results using the method of least squares [169, 170] for the temperature dependence of density gave the following equations:

$$\rho_{Sn} = (7001 \pm b) - (0.7 \pm 0.01)(T - 505)\ \text{kg/m}^3 \qquad (4.1.1)$$

$$\rho^1_{In} = (7040 \pm 37) - (0.79 \pm 0.12)(T - 425)\ \text{kg/m}^3 \qquad (4.1.2)$$

$$\rho^2_{In} = (7074 \pm 24) - (0.83 \pm 0.07)(T - 425)\ \text{kg/m}^3 \qquad (4.1.3)$$

$$\rho_{Pb} = (10696 \pm 10) - (1.18 \pm 0.12)(T - 605)\ \text{kg/m}^3 \qquad (4.1.4)$$

$$\rho_{Tl} = (11150 \pm 29) - (1.06 \pm 0.23)(T - 575)\ \text{kg/m}^3 \qquad (4.1.5)$$

$$\rho_{Ga} = (6089 \pm 8) - (0.64 \pm 0.03)(T - 300)\ \text{kg/m}^3 \qquad (4.1.6)$$

$$\rho^1_{Bi} = (9946 \pm 22) - (1.16 \pm 0.18)(T - 575)\ \text{kg/m}^3 \qquad (4.1.7)$$

$$\rho^2_{Bi} = (9957 \pm 1) - (1.25 \pm 0.01)(T - 575)\ \text{kg/m}^3 \qquad (4.1.8)$$

The measurements of the density of indium and bismuth were taken in several experimental series using different areometers. Comparison of the experimental data shows (see the equations (4.1.2), (4.1.3) and (4.1.7), (4.1.8)) that the differences in the experimental data are the range of the permissible error of measurement of density by the areometer. Our results are close to the values recommended in [171, 172].

The temperature dependence of the surface tension of metals (with the exception of gallium) was also linear with negative values of the temperature coefficient. In the experimental range, the resultant data coincide with the most reliable literature data [12] and approximated by linear functions:

$$\sigma_{Sn} = (542 \pm 2) - (0.068 \pm 0.004)(T - 505)\ \text{mJ/m}^2 \qquad (4.1.9)$$

$$\sigma_{In} = (558 \pm 3) - (0.079 \pm 0.008)(T-425) \text{ mJ/m}^2 \quad (4.1.10)$$

$$\sigma_{Pb} = (444 \pm 2) - (0.080 \pm 0.018)(T-605) \text{ mJ/m}^2 \quad (4.1.11)$$

$$\sigma_{Bi} = (376 \pm 2) - (0.064 \pm 0.009)(T-575) \text{ mJ/m}^2 \quad (4.1.12)$$

$$\sigma_{Tl} = (468 \pm 0.3) - (0.094 \pm 0.002)(T-575) \text{ mJ/m}^2 \quad (4.1.13)$$

The temperature dependence of the surface tension of gallium was determined several times in different devices. The first experimental series showed deviations of the polytherms from linearity: in the initial section from the melting point to 423 K we observed temperature buffering, *i.e.*, tension did not change with temperature. With a further increase of temperature, the tension monotonically decreased. Mathematical processing of the results gave the following equation:

$$\sigma = 707 - 1.2\, 10^{-3}\,(T-303) - 9.0 \cdot 10^{-5}\,(T-303)^2 \quad (4.1.14)$$

The non-linear dependence of the surface tension and temperature was also determined by other authors [173–177]. The authors of [174] compared the phenomenon with the increase of the coordination number of gallium in a temperature range above the melting point [188]. In [177] gallium of higher purity was used and the linear dependence of surface tension on temperature was obtained.

In experiments, we used high purity metals and the reason for nonlinearity should therefore be found elsewhere. The surface tension of metals is highly sensitive to impurities [180] and, consequently, to the experimental conditions. We increased requirements on the preliminary preparation of equipment for measurements. In addition to the procedures described in the previous chapter, the entire system was subjected (prior to charging) to long-term (approximately 30 h) thermal vacuum treatment. The device was heated to 793 K, and the loading hoppers with the metals up to 1073 K. The loading cylinders were prepared especially for this purpose from quartz glass. The measurements, taken by this procedure, show that the temperature dependence of the surface tension of pure gallium is linear with a negative temperature coefficient:

$$\sigma_{Ga} = 719 - 0.061(T-303) \text{ mJ/m}^2$$

In the second chapter, in the model of the homogeneous layer with a finite thickness, we derived equations for calculating the temperature dependence of the molar area and the effective thickness of the surface layer of pure substances. For these calculations, it is necessary to have also data on the temperature dependence of the surface tension and density.

Using the experimental results, we calculated the temperature dependence of the thickness of the surface layer and the molar area of the investigated metals. The values of the physical–chemical properties, required for calculations, were taken from [183–186]. In the calculations, it was assumed that at the melting point the density of the surface layer is 95% of the density of the liquid phase. This *apriori* assumption reduces to a certain degree the value of the results, nevertheless, these results make it possible to calculate the values of the thickness of the surface layer of different metals and examine its temperature dependence.

The data are presented in Table 1, which shows that the thickness of the surface layer of tin and gallium is close to the monolayer, and that of lead, bismuth and thallium to two monatomic layer. In the temperature range from the liquidus temperature to 773 K the thickness of the surface layer of the investigated metals depends only slightly on temperature. This results in agreement with

Table 1. Effective thickness of the surface layer and molar area of pure metals

	T, K	473	573	673	773
Indium	$\omega \cdot 10^{-4}$ (m²/mol)	3.65	3.67	3.69	3.70
	n (number of monolayers)	1.5		1.5	1.5
Tin	$\omega \cdot 10^{-4}$ (m²/mol)	-	5.60	5.64	5.68
	n (number of monolayers)	-	1.0	1.0	1.0
Gallium	$\omega \cdot 10^{-4}$ (m²/mol)	4.31	4.34	4.35	4.36
	n (number of monolayers)	1.1	1.1	1.1	1.1
Thallium	$\omega \cdot 10^{-4}$ (m²/mol)	–	–	2.81	2.82
	n (number of monolayers)	–	–	2.1	2.1
Lead	$\omega \cdot 10^{-4}$ (m²/mol)	–	–	3.46	3.47
	n (number of monolayers)	–	–	1.8	1.8
Bismuth	$\omega \cdot 10^{-4}$ (m²/mol)	–	3.92	3.93	3.96
	n (number of monolayers)	–	1.6	1.6	1.7

the conclusions in [68] according to which the thickness of the surface layer of liquid metals is several monomolecular layers. These results are also consistent with the Rusanov criterion.

It is believed [184] that because of the small radius of the effect of molecular forces, the surface layer of the metals cannot exceed one or two monatomic layers. However, the authors of [185] show that in the metals, the potential of the effective interaction is oscillating and acts over a longer distance than the Lennard–Jones potential. This shows that several monatomic layers may be included on the surface layer, and this is also confirmed by the calculations carried out by the authors of [186–187].

Thus, the equations presented in chapter 2 may be used to calculate the effective thickness and molar areas of pure metals using the same model, without any need for data on the properties of binary multicomponent systems. Consequently, the concentration dependences of the discussed properties in melts may be determined. The results of these calculations are discussed below.

4.2. Surface tension and density of indium–tin and gallium–lead binary systems

The literature data on the surface tension and density of melts of the indium–tin system are contradicting [190–192]. Therefore, we measured the values of σ and ρ of 25 melts of this system. The temperature dependence of the density of all investigated alloys was linear and approximated by the equation:

$$\rho=\rho_0 - (T - T_m)$$

Here ρ_0 and α are the physical constants whose values depend on the composition of the alloy. The values of these quantities and of the appropriate confidence ranges are presented in Table 1 of the appendix.

The density data were used in calculating the molar volumes. The isotherms of the molar volumes, plotted on the basis of the results of these calculations, characterised by small positive deviations from the additive straight lines (Fig. 4.1). This indicates that the formation of melts is accompanied by an increase of the volume. The maximum deviations were recorded in the range of medium compositions and equal 1% at 523 K. With the increase of temperature, the values of the molar volumes approached the additive values and at 773 K these deviations slightly exceed the experimental error.

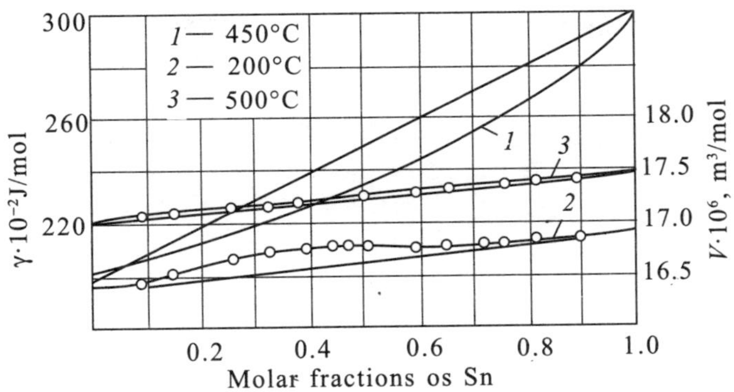

Figure 4.1. Isotherms of molar tension (1) and molar volumes (2, 3) in the indium–tin system.

All the investigated alloys are characterised by a decrease of surface tension with increasing temperature. The isotherms σ of the majority of melts are described by linear equations. Only in certain alloys, enriched with indium, there are small deviations of the temperature dependence of surface tension from linear. Although the errors of the linear approximation of these results are located in the error range of the measurements, nevertheless, the noted special feature is detected in all alloys with a high indium content. Therefore, in addition to the coefficients of the approximating equation, the Appendix gives the experimental data on the surface tension of the investigated alloys (Tables 2 and 3).

The concentration dependence of the temperature coefficient of surface tension $\left(\frac{d\sigma}{dT}\right)$ is not simple. With the increase of the tin content, the absolute value of the temperature coefficient decreases and reaches the minimum value at mean compositions. A further increase of the tin content results in a monotonic increase of the coefficient to the value of the temperature coefficient of surface tension of pure tin.

The concentration dependences of the surface tension of the investigated system at 523, 623 and 773 K are presented in Fig. 4.2. The most interesting special feature of these isotherms is the flat minimum in the range of mean composition. As tin on indium, indium is also surface active on tin. However, the limiting surface activity of tin on indium is considerably higher than that of indium on tin.

Comparative analysis of the isotherms, constructed at different temperatures, showed the following special features. As expected,

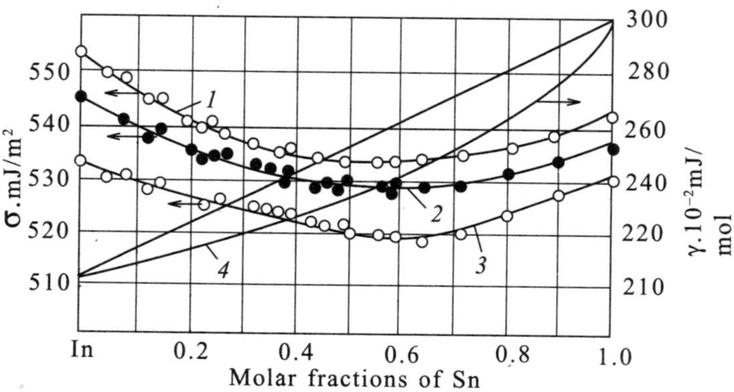

Figure 4.2. Isotherms of surface tension at 523, 623 and 773 K in the indium–tin system.

the increase of temperature results in a decrease of the depth of the minimum. This is accompanied by a small displacement of the minimum from the region of mean compositions. A decrease of the depth of the minimum with increasing temperature of the melts enriched with indium is much faster than in the melts enriched with tin. This is caused by the rapid decrease with temperature of the limiting surface activity of tin and a weak dependence of this parameter in indium.

The literature data on σ and ρ of indium–tin melts differ not only quantitatively but also quantitatively. The density of the melts of this system was investigated in [190–192]. In [192], positive deviations of densities from additive values were obtained, in [191] the deviations were negative, and an additive dependence was obtained in [190]

The surface tension of these alloys was measured by the authors of [189–191]. In [190] smooth σ curves, bent in the direction of the composition axis, were obtained for the concentration dependence of σ. However, the authors of [189, 191] showed that the isotherms of σ in this system are characterised by a flat minimum in the range of mean compositions.

The results of measurements are in good agreement with the data in [189, 191].

Taking these considerations into account, it is interesting to examine the literature data on the surface tension and density of the thallium–lead binary system which in the thermodynamic aspect is very similar to the indium–tin melts. The surface tension of the thallium–lead melts was examined by the gravitational method proposed by Rusanov from the liquidus temperature to 773 K [191] and by the sessile droplet method up to 1273 K [193]. The data on the surface tension, obtained by different methods, are in good agreement and indicate that the isotherms of σ are characterised by flat minima in the range of mean composition, as in the indium–tin system.

The density and molar volume of these melts were investigated in [191, 193, 194–196]. The authors of [191, 193] obtained positive deviations of the molar volumes from additivity, and in [195] the authors presented a more complicated isotherm of the molar volumes, in particular, the isotherms of partial–molar volumes of components in this system according to the data in [195] proved to be sign-changing.

We have taken repeated measurements of the concentration dependence of the surface tension of this system. The results [111] are presented in Table 4 of the Appendix and are in qualitative agreement with the data published in [191, 193]. It is important to note the most interesting result: the temperature dependence of the surface tension of the investigated alloys in the given temperature range is linear. The values of the temperature coefficient are situated between the values of $\left(\frac{d\sigma}{dT}\right)$ of pure metals.

The isotherms of surface tension (Fig. 4.3) show, as in the indium–tin system, minima at 0.4 molar fractions of thallium. With the increase of temperature, the depth of the minimum slightly decreases but its abscissa remains unchanged. The isotherms of the molar volumes indicate the alternating nature of the volume effects

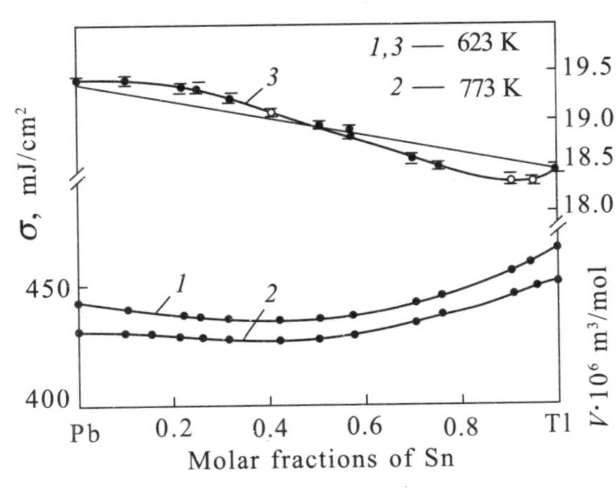

Figure 4.3. Isotherms of surface tension and molar volumes in the thallium–lead system.

of mixing of the melts. Formation of the alloys, enriched with lead, is accompanied by an increase of the volume in comparison with the additive value, and starting with the equimolar composition, the volume effect has the reverse sign. As mentioned previously, the alternating nature of the partial–molar volumes in this system has also been found by the authors of [195].

Thus, the indium–tin and thallium–lead binary systems have similar surface tension isotherms. The melts of the systems are also similar as regards the thermodynamic aspect, as indicated by the data on the physical–chemical properties [195–197].

The authors of [189] calculated the isotherms of surface tension of indium–tin and thallium--lead binary systems using the available theoretical equations [80, 63, 92, 199, 200]. The results of the calculation show that these equations do not express the dependence of the surface tension on composition in these systems.

The physical interpretation of the special features of the surface tension isotherms is not possible without reliable data on the main parameters of the surface layer of the melts: adsorption of components, composition and effective thickness of the surface layer, molar and partial molar surfaces, etc. Therefore, the results of calculation of these properties are presented below.

4.3. Special features of the concentration dependence of the thermodynamic parameters of the surface layer of indium–tin and thallium–lead melts

Using the procedure, proposed in the second chapter, we calculated the adsorption (N-variant according to Guggenheim and Adam), composition and effective thickness, molar and partial molar surfaces of the component in the investigated systems.

The adsorption isotherms of tin at 673 and 773 K are shown in Fig. 4.4. The concentration dependence of the adsorption of tin is complicated. In the range of low concentrations of tin, the adsorption of tin is positive and at 673 K reaches the maximum value ($0.72 \cdot 10^{-6}$ mol/m^2) in the alloy with the molar fraction of tin of 0.25. A further increase of the tin content of the melt results in a smooth decrease of the adsorption isotherm which passes through the minimum value ($-0.75 \cdot 10^{-6}$ mol/m^2) at 0.8 molar fractions of tin.

With the increase of temperature the absolute value of adsorption decreases. This is accompanied by a small displacement of the

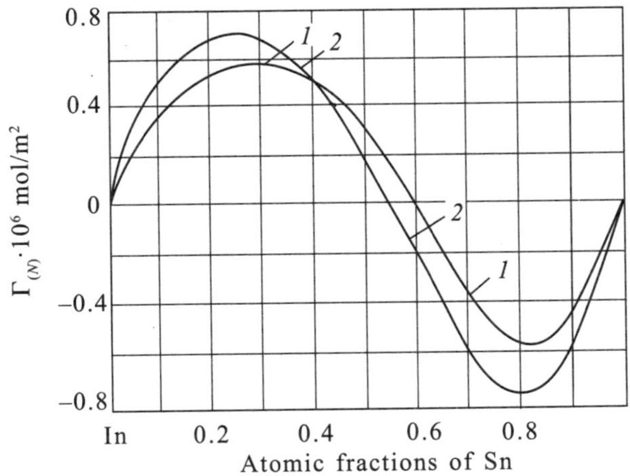

Figure 4.4. Adsorption isotherms of tin in the indium–tin system at 673 and 773 K.

extreme points (maximum, minimum) in the direction of tin. This is caused by the displacement of the minimum on the surface tension isotherms in the direction of tin.

The adsorption phenomena in thallium–lead melts have been examined in detail by the authors of [191, 193]. Our results are in agreement, within the experimental error range, with these data. Therefore, we confine ourselves to stating that the adsorption phenomena in the indium–tin and thallium–bismuth binary systems have many common features.

As indicated by Fig. 4.5 and 4.6, the dependence of the effective thickness on the composition of the surface layer in the indium–lead and thallium–lead systems is the same. The increase of the content of the component with a smaller thickness of the surface layer (tin or lead) results in a monotonic decrease of the thickness of the surface layer of the alloy. The isotherms of the thickness of the surface rainbow systems showed the inflection points in the range of mean composition.

The form of the curves of the isotherms molar fractions of the component on the surface layer is also the same. These isotherms are shown in Fig. 4.3. Isotherms of surface tension and molar volumes in the thallium–lead system are S-shaped, indicating that the composition of the surface layer in the entire concentration range is close to equimolar, in comparison with the composition of the melt.

Thus, the concentration dependence of the thermodynamic parameters of the surface layer of the indium–tin and thallium–lead

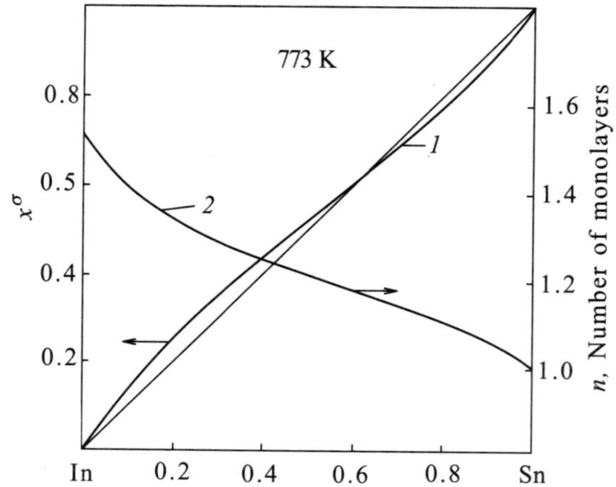

Figure 4.5. Isotherms of molar fractions of tin in the surface layer (1) and the dependence of thickness on the composition of the surface layer (2) in the indium–tin system at 773 K.

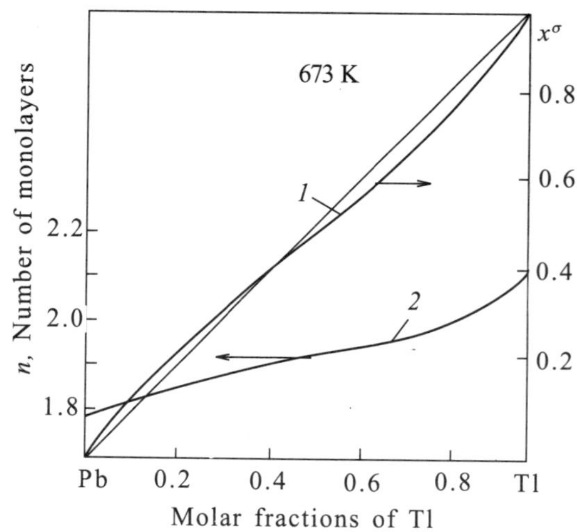

Figure 4.6. The composition and thickness of the surface layer of the lead–thallium melts at 623 K. 1) the thallium content of the surface layer; 2) the dependence of thickness on the composition of the surface layer.

binary systems has the same form. Under physical–chemical properties of these melts are also similar. The most interesting feature is the presence of a minimum on the isotherms of surface tension. The similarity of the discussed melts in the thermodynamic respect makes it possible to assume that the nature of the extrema on the surface tension isotherms is the same.

In the literature, various assumptions have been made regarding the reasons for the formation of extrema on the surface tension isotherms [142, 201–204). In many cases, these special features are attributed to structural changes in the melts, such as the formation of a certain number of large aggregates of particles or dipole pairs, and also to the variation of the interaction of the particles in the surface layer in comparison with the interaction in the volume of the melt. At similar values of the surface tension of the pure components, any of the reasons which may slightly reduce the free surface energy, made lead to the formation of a minimum on the surface tension isotherms.

Of the currently available literature data, special attention should be given to the results of experimental examination of the composition of the surface layer of indium–tin melts by Auger electron spectroscopy [204]. In this study it was established that the purification of the surface of a metal or alloy to remove carbon in the oxygen medium changes qualitatively the form of the surface tension isotherms. Using the data on the composition of the surface layer, obtained by Auger spectroscopy, the authors of [204] calculated, from the Gibbs adsorption equation, the surface tension of the indium–tin melts and obtained the isotherm with a broad maximum, instead of a minimum. This is in agreement with the predictions of the simulation–thermodynamic theories of surface tension [92].

However, as shown in chapter 3, the experimental equipment and measurement procedure which we developed prevent the penetration of large amounts of carbon into the system. Therefore, we interpret these special features differently.

The indium–tin and thallium–lead binary systems are close to ideal solutions. The pure components, indium, tin, thallium and lead are characterised by similar values of the physical–chemical properties: the size of the atoms, the melting and boiling points, vapour pressure, surface tension, density, molar volumes, and the melts of these systems are characterised by small sign-changing deviations of the activity coefficient from unity. The sign-changing deviations from additivity are also detected on the isotherms of molar volumes.

It is well known [205, 206] that the ideal solutions may form component cells atoms have similar dimensions and structure. The main condition of the formation of the ideal solution is the equality of the potentials of intermolecular interaction for all molecules (atoms A–A, A–B). Evidently, these conditions are approximately fulfilled in the discussed systems.

On the basis of a small but mostly positive deviation of the molar volumes and heats of mixing from additivity, it may be assumed that the energy of interaction of different atoms in the systems is lower than that of the atoms of the same type. Therefore, taking into account the requirement of the minimum free surface energy, in the condition of thermodynamic equilibrium the most suitable composition of the surface layer and the distribution of the atoms interlayer is the one which ensures the preferential combination of the bonds of different atoms. However, the ordered distribution of the atoms in the surface layer is counteracted by the temperature factor whose effect becomes greater with increasing temperature.

With similar dimensions and properties of the indium and tin atoms, the maximum number of the bonds of the different atoms is possible at the equimolar composition of the surface layer. This may be used to explain the fact that the composition of the surface layer of all indium–tin and thallium–lead alloys is closer to equimolar than the composition of the volume of the melt. Consequently, at low temperatures, when the temperature factor does not play the controlling role, the lowest value of the surface tension should be recorded in the alloy with the equimolar composition.

An increase of temperature increases the importance of the role of the entropy factor. This results in the disordering of the atoms in the surface layer and, consequently, in a decrease of the number of bonds of different atoms. For the constant composition of the surface layer of issue results in a relative increase of the surface tension of the melt, but this is counteracted by adsorption processes.

At the constant number of bonds of the different atoms, the adsorption of tin is more advantageous in the energy aspect in the group of the pure components, because the surface tension of tin is lower than that of indium. Actually, the calculations describes previously, confirmed that the extent of adsorption of tin in the alloy with the equimolar composition increases with increasing temperature. Consequently, an increase of temperature results in positive adsorption of tin in the alloy with the equimolar composition and the point, corresponding to zero adsorption or the minimum on the surface tension isotherms, is displaced in the direction of tin.

These considerations show that the special features (minimum, maximum) may be found not only in real but also in ideal solutions, governed by the Raoult's law. It is necessary to determine whether this conclusion contradicts the laws of thermodynamics. The ideal surface tension isotherms, obtained from the well-known equations, smooth curves, now at bent in the direction of the composition axis.

This question will be examined in greater detail.

The Gibbs adsorption equation shows that if the surface tension isotherm of the binary system is characterised by the extreme points, then irrespective of the method of determination, the adsorption of components converts to zero, *i.e.*

$$\Gamma_{2(1)} = \Gamma_{1(2)} = \Gamma_{2(P)} = 0 \tag{4.3.1}$$

Instead of the adsorption, we introduce its value:

$$\Gamma_1 = \int_{-\infty}^{0}\left[n_1(z) - n_1^{\alpha}\right]dz + \int_{0}^{+\infty}\left[n_1(z) - n_1^{\beta}\right]dz = 0 \tag{4.3.2}$$

$$\Gamma_2 = \int_{-\infty}^{0}\left[n_2(z) - n_2^{\alpha}\right]dz + \int_{0}^{+\infty}\left[n_2(z) - n_2^{\beta}\right]dz = 0 \tag{4.3.3}$$

The determination of the general conditions, essential for the simultaneous fulfilment of equations (4.3.2), (4.3.6), is a complicated and difficult to solve task However, if the profiles of the local concentration of the components are linked by the equation $n_i(z) = an^2(z)$ (a = const), then irrespective of the type of functions $n^i(z)$, the integrals (4.3.2) and (4.3.3) for the same integration ranges convert simultaneously to zero. In a partial case at $a = 1$, $n^1(z) = n^2(z)$, *i.e.*, the profiles of the local concentration of the component are identical. It may easily be seen that this behaviour of the profiles of local concentrations is possible only in the ideal solutions with the equimolar composition, because these solutions are described by the equalities: $n_1^{\alpha} = n_2^{\alpha}$; $n_1^{\beta} = n_2^{\beta}$.

Thus, the claim regarding the possibility of formation of a minimum on the surface tension isotherms of ideal solutions does not contradict the laws of thermodynamics. Consequently, one of the possible reasons for the formation of a minimum on the surface tension isotherms may be the identical form of the profiles of the local concentrations of components. This means that it may be carried out for the same dependence of the local concentrations of the components on the coordinate z, normal to the fracture surface.

In conclusion, it should be mentioned that these conditions are sufficient but not essential for the appearance of an extremum. There are many possible reasons for the formation of extrema. It has been shown previously that in the specific conditions, namely at $n_1(z) = \alpha n_2(z)$, the surface tension isotherms of ideal solutions may show an extremum in the form of a minimum or maximum.

4.4. Surface properties of binary systems: bismuth–thallium, tin–thallium and indium–thallium

The surface tension and density of indium–thallium and tin–thallium binary systems have been determined by the authors of [189, 192, 207, 208]. The surface tension isotherms of these melts contain flat minima which could not be explained on the basis of the structure of the melt [195] and contradicts the theoretical predictions [209]. On the other hand, examination of the surface properties of indium–tin–thallium and indium–bismuth–thallium ternary systems showed discrepancy between the experimental data for the ternary melts and these melts.

The surface properties and density of the thallium–bismuth melts have not been studied extensively. We know of one study [208] in which a surface tension isotherm was generated at 673 K, together with examination of electrocapillary phenomena. However, we know of no data on the temperature dependence of the surface tension and density of these melts.

Because of the situation, we have examined the temperature concentration dependence of the surface tension and density of these systems.

The results are presented in the tables 5–10 of the Appendix. The temperature dependence of the surface tension and density of the melt of all the investigated systems is linear with a negative temperature coefficient.

The concentration dependence of the temperature coefficient of surface tension for all the investigated systems has many common features. In particular, the increase of the content of the surface-active component (bismuth in the thallium–bismuth system, thallium in the indium–thallium and tin–thallium systems) to the equimolar composition results in a larger decrease of the temperature coefficient (the absolute value). With a further increase of the content of the second component, its value approaches monotonically the temperature coefficient of the surface tension of the second component.

The concentration dependence of the surface tension and the molar volumes in the indium–thallium and tin–thallium binary systems is more or less identical. In particular, the isotherms of the molar volume in both systems coincide with the additive straight lines. As an example, Fig. 4.7 shows the isotherms of molar and partial molar quantities in the thallium–indium system. This form of

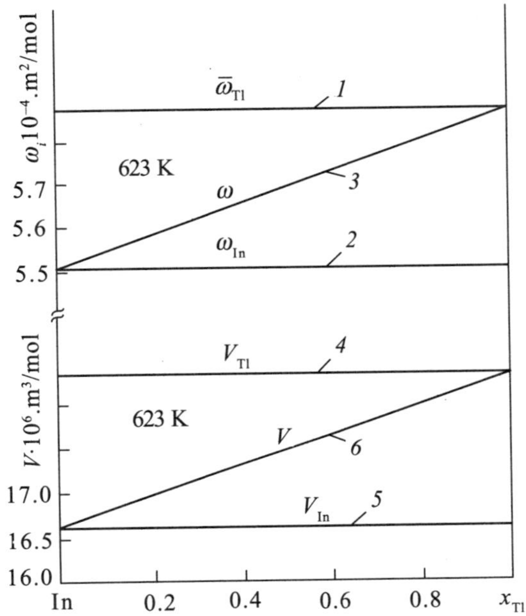

Figure 4.7. The isotherms of the molar and partial–molar quantities (volume and surface) in the indium–thallium system at 623 K.

the relationship is also characteristic of the tin–thallium melts. It should be mentioned that the molar surfaces have been calculated for the monomolecular surface layer, and the partial molar surfaces both from the well-known approximate equation (2.2.8) and the more accurate equation (2.2.14). The calculation show that the difference in the data is, as expected from comparative analysis of these relationships (2.6.7), within the error range of determination of the partial molar quantities.

Figures 4.8 and 4.9 show the surface tension isotherms and density of the indium–thallium and tin–thallium binary systems, respectively. The dependence of the density on the molar fraction of thallium in the investigated systems is described by an additive function.

The concentration dependence of the surface tension in both systems is expressed by smooth curves, bent in the direction of the composition axis. Thallium is a surface-active element on both indium and tin. However, the limiting surface activity of thallium in indium is slightly lower in the absolute value than in tin. In the melts of the investigated systems, diluted with respect to thallium, the increase of the thallium content is accompanied by a rapid decrease

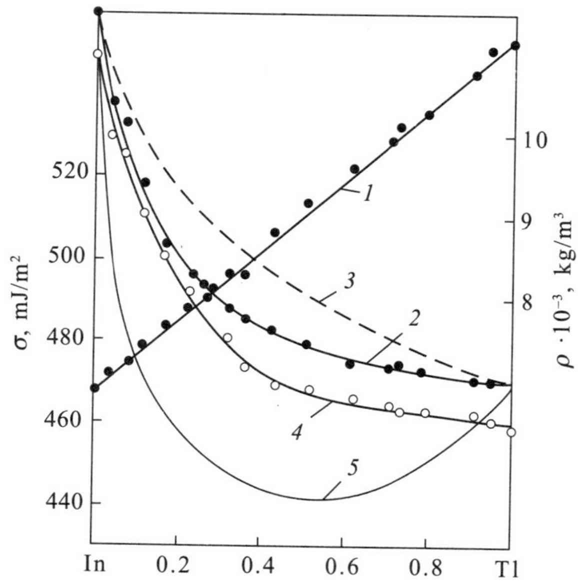

Figure 4.8. Surface tension and density of indium–thallium melt. 1) the density isotherm at 623 K; 2, 4) the surface tension isotherms at 623 and 773 K, respectively; 3) the data obtained from the Popel'–Pavlov equation at 623 K; 5) the literature data.

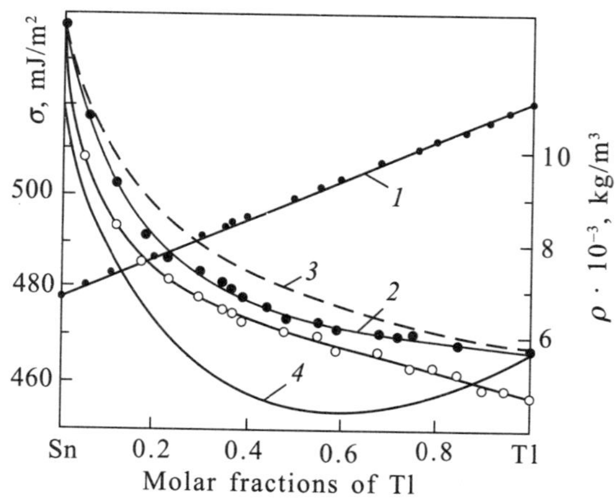

Figure 4.9. Surface tension and density of tin–thallium melts. 1) density isotherm at 623 K; 2) the surface tension isotherms at 623 K; 3) the values obtained from the Popel'–Pavlov equation at 623 K.

of surface tension. A further increase of the values of the molar fraction of thallium results in a monotonic decrease of the surface tension to the value of σ for pure thallium.

For comparison, the figures also show theoretical isotherms. In the calculations using theoretical equations, we used our data on the partial molar surfaces, and also the literature data on the

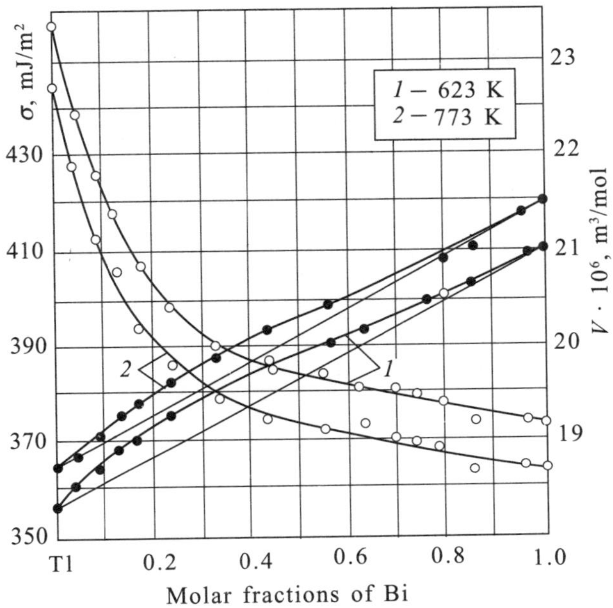

Figure 4.10. Isotherms of the surface layer and molar volumes of the thallium–bismuth system: 1) 623 K; 2) 773 K.

thermodynamic activities of the component [212]. The calculations show that the Popel'–Pavlov equation [199] describes with sufficient accuracy (the difference is smaller than 3%) the concentration dependence of surface tension in the indium–thallium and tin–thallium binary systems.

The concentration dependence of the surface tension in the thallium–bismuth system (Fig. 4.10) does not differ qualitatively from the surface tension isotherms of the indium–thallium and tin–thallium binary melts. At all temperatures, the surface activity of bismuth on thallium is high. With the increase of temperature the limiting surface activity of bismuth on thallium decreases.

The values of the density of the melts were used to calculate the molar volumes. The results of the calculations show that the formation of the thallium–bismuth melts is accompanied by an increase of the volume. The maximum deviations of the isotherms of the molar volumes from the additive curves are recorded in the range of the medium compositions and at 626 K they equal 2% which is slightly higher than the measurement error.

Thus, our experimental data show the absence of the minima on the isotherms of surface tension of the indium–thallium and tin–thallium binary systems. The isotherms have the form of smooth curves described satisfactorily by the Popel'–Pavlov simulation–thermodynamic equation.

The experimental data on the surface tension and density may be used to determine the adsorption of the components, the distance between the individual positions of the dividing surface, the composition and thickness of the surface layer and the molar and partial–molar surfaces. These calculations were carried out using the procedure described in chapters 1 and 2. It should be mentioned that the distance between the Gibbs ($\Gamma_{Tl} = 0$) and equimolar $\left(\sum_{i=1}^{n}\Gamma_{i(n)} = 0\right)$ dividing surfaces is calculated from the equation:

$$Z_{Tl-N} = -\frac{1}{n_{Tl}}\Gamma_{Tl(n)}$$

where $\Gamma_{Tl}(n)$ is the adsorption on the equimolar dividing surface (N – the variant according to Guggenheim and Adam), n_{Tl} is the number of atoms in the unit volume of the melt.

The indium–thallium, tin–thallium and thallium–bismuth melts do not differ greatly from the ideal state and their surface tension isotherms are similar. This determines the identical form of the adsorption isotherms of surface-active components in these melt. In fact, Figs. 4.11 and 4.12 show that the adsorption isotherms of

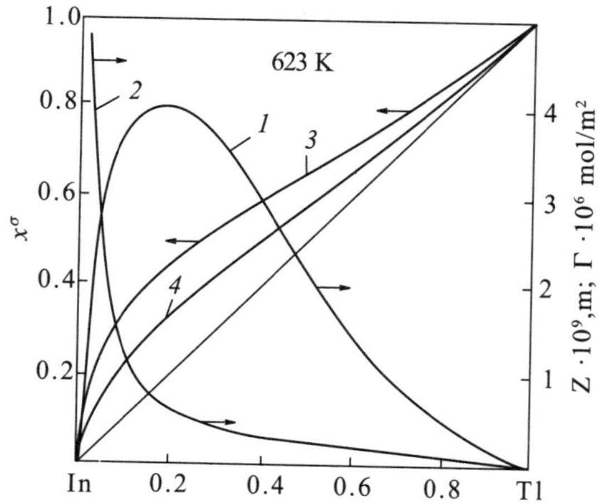

Figure 4.11. Adsorption, composition of the surface layer and the distance between the different positions of the dividing surface in the indium–thallium system at 623 K. 1, 2) adsorption isotherms of bismuth at 523 and 773 K; 3) the molar fractions of bismuth at 773 K; 4) the distance between different positions of the Gibbs dividing surface.

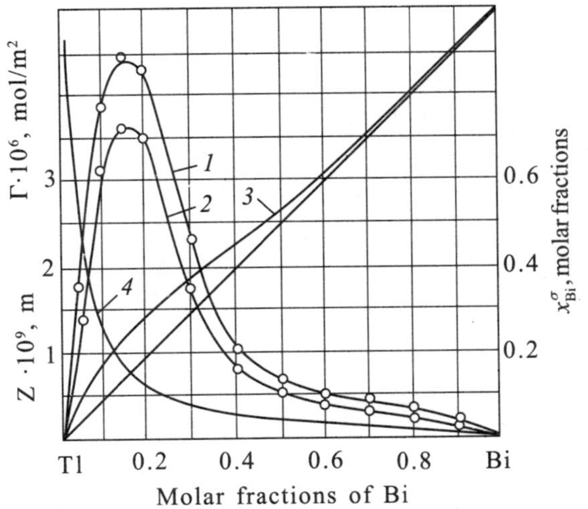

Figure 4.12. Isotherms of parameters of the surface layer of the thallium–bismuth melts. 1) adsorption of thallium; 3, 4) the isotherms of molar fractions of thallium in the surface layer, calculated for the monolayer and using the procedure described in chapter 2; 2) the distance between the Gibbs dividing surfaces.

thallium in the indium–thallium and tin–thallium melts and of bismuth in the thallium–bismuth system have the same form. In all the melts of the indium–thallium and tin–thallium systems, thallium is adsorbed positively. The maximum value of thallium adsorption is obtained at 0.1 and 0.2 molar fraction of the thallium in the tin–thallium and indium–thallium systems, respectively. A further increase of the thallium content of the melts results in a monotonic decrease of thallium adsorption to zero. The behaviour of the adsorption isotherms of bismuth in the Guardian–bismuth system is identical (Fig. 4.12).

Figures 4.11 and 4.12 also showed the distances between the different positions of the dividing surface. In both systems at high thallium concentrations, Z_{Ti-N} has the order of atomic spacing. However, with a decrease of the thallium content, the value Z_{Tl-N} increases in the absolute value and in the limit at $X_{Tl} \to 0$ tends to infinity. This means that one of the investigated dividing surfaces differs from the physical boundary between the phases.

The problem of advantages of methods of determination of adsorption has been discussed quite often in the literature. The main advantage of adsorption is the independence of the position of the dividing surface on the composition of the melt and its proximity to the physical interface. Therefore, this problem will be examined in greater detail.

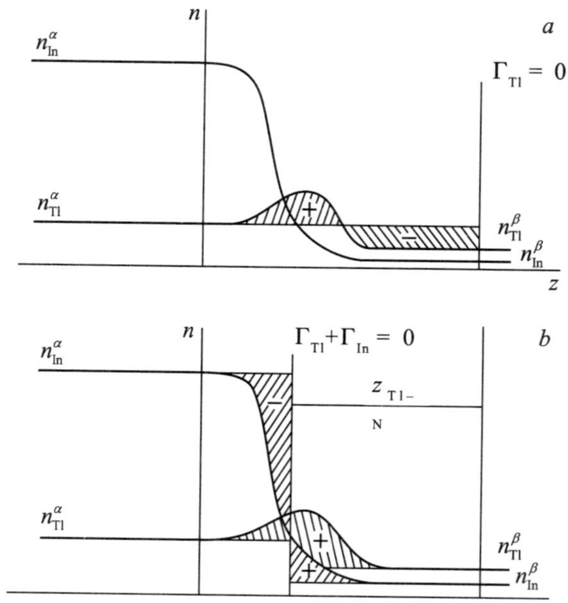

Figure 4.13. Profiles of local concentrations in melts and different positions of the dividing surface. a) the Gibbs dividing surface, b) the equimolar dividing surface.

To provide more information, Fig. 4.13 shows the possible profiles of the local concentrations in a solution diluted with respect to thallium. Different positions of the dividing surface are also indicated. The figure shows that with the decrease of the thallium content, the distance of the Gibbs dividing surface from the physical interface increases, and the position of the dividing surface is given by the condition ($\Gamma_{Tl} = 0$). The position of the equimolar dividing surface changes only slightly.

Analysis of the graphs shows that the conclusions regarding the concentration dependence of $Z_{Tl—N}$ are generally valid and hold for any profile of the local concentration. Consequently, the position of the Gibbs dividing surface greatly changes with the composition of the melt. However, this fact does not indicate that the Gibbs variant is not suitable for the analysis of adsorption phenomena. It has been shown previously that all the methods of determination of adsorption are equally important and the Gibbs dividing surface is preferred in a number of cases.

Using the experimental data determined using the procedure described previously, we calculated the surface properties of indium–thallium and tin–thallium binary melts. The isotherms of these properties in both systems do not differ qualitatively.

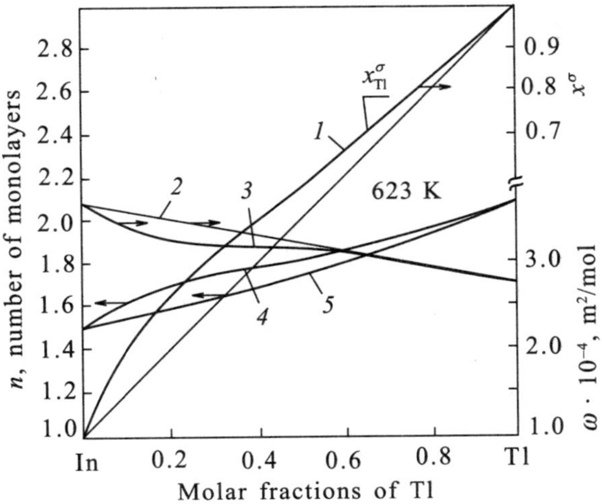

Figure 4.14. Isotherms of surface properties of indium–thallium melts at 623 K. 1) the molar fractions of thallium in the surface layer; 2, 3) the dependence of the molar area on the composition of the surface layer and the solution, respectively; 4, 5) the dependence of the thickness of the surface layer on the thallium content of the solution and in the surface layer, respectively.

Therefore, Fig. 4.14 shows the graphs only for the indium–thallium binary system. Evidently, the conclusions and relationships governing the variation of these properties, determined for the system, also hold for the tin–thallium melts.

The values of the thickness of the surface layer of pure metals, required for calculations, were determined in section 4.1. Curve 1 in Fig. 4.14 shows that the surface layer is enriched with thallium atoms in the entire concentration range. The maximum difference between the composition of the surface layer and the melt is obtained at 0.2 molar fractions of thallium.

The isotherms 2 and 5 indicate that the dependences of the molar surface and thickness on the composition of the surface layer are transferred by the additive function. However, the dependences of these properties on the composition of the solution (isotherms 3 and 4) greatly deviate from the additive straight lines. If the isotherms of the molar surfaces are characterised by the negative deviations from additivity, the dependence of the thickness of the surface layer on the composition of the solution has a positive deviation from the additive straight line. The special features are explained by the adsorption processes on the surface of the melts.

In fact, the considerably faster increase of the thickness of the surface layer in the initial section of the isotherms 4 is caused by the positive adsorption of thallium in diluted indium–thallium and tin–thallium solutions. It should be taken into account that the thickness of the surface layer of thallium is greater than that of indium or tin. The same explanation may be used for the negative deviations of the molar surfaces from additivity.

4.5. Surface tension and molar volumes in gallium–bismuth melts

A characteristic feature of gallium systems is the complicated form of the isotherms and polytherms of surface tension [12, 59, 190, 123, 214]. Evidently, this is the reason for the considerable attention paid to the examination of the properties of gallium systems. Our results obtained for ternary (In–Sn–Ga) and quaternary (In–Sn–Ga–Bi) melts are in good agreement with the data on the surface tension and density of indium–gallium [213], tin–bismuth [219], indium–bismuth [220] and indium–tin [191] binary systems. However, the experimental data on the multicomponent melts differ from the literature data on the concentration dependence of surface tension in the gallium–bismuth binary system [213]. It should be mentioned that investigations of the gallium–bismuth and gallium–tin [213] binary systems yielded complicated isotherms with a minimum and a maximum (Fig. 4.14) which could not be interpreted on the basis of the existing theoretical models. Therefore, we have again examined the gallium–bismuth binary system. Careful verification of the experimental data shows that special features of the isotherms of σ of the gallium–tin binary system, discussed in [218], fit in the range of the error and do not require any further attention.

We have also examined the temperature dependence of the surface tension and density of 18 alloys of the gallium–bismuth binary system.

The measurements of the surface tension were taken using two methods: the maximum pressure in the droplet (3.5) in the temperature range 573–773 K, and by the method of the large droplet at 573–773 K in a device described in [215]. Measurements in the combined device were taken using the procedure described in chapter 3. Investigations by the large drop the methods were carried out in helium purified by passage through activated coal at the liquid nitrogen temperature. Substrates were produced from grade TG-1500 graphite. The upper edge of the substrate was sharpened. The results are presented in the Tables 9 and 10 of the Appendix.

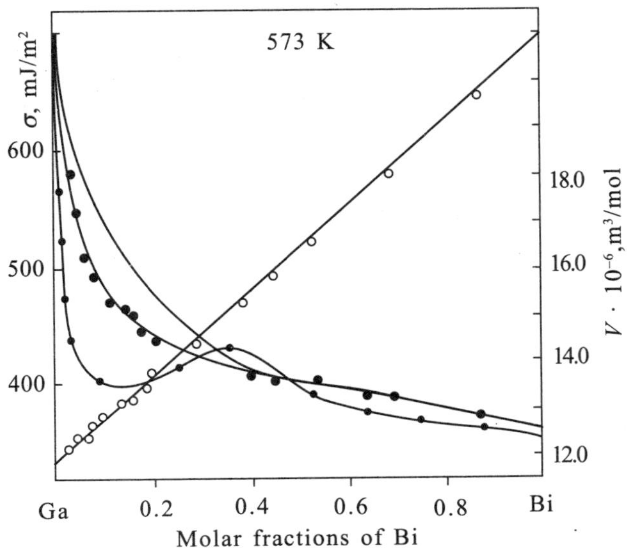

Figure 4.15. Isotherms of the molar volumes and surface tension of the gallium–bismuth system at 573 K;1) the molar volumes; 2) surface tension isotherms; 3) literature data; 4) the isotherms calculated from the Popel'–Pavlov equation.

As expected, the temperature dependence of the density of the gallium–bismuth melts is linear with a negative temperature coefficient.

The isotherms of the molar volumes (Fig. 4.15) coincide with the additive straight lines was in the limits of the error range. The gallium–bismuth melts are characterised by a positive deviation from Raoult's law, and their formation should be accompanied by an increase of the volume. However, as shown in [216], the investigated system should show a compression of the volume determined by the large difference in the dimensions of the atoms (the molar volume of bismuth is $21 \cdot 10^{-6}$ m³/mole, and that of gallium $12 \cdot 10^{-6}$ m³/mole). Consequently, the volume effect of the formation of the gallium–bismuth alloys may be compensated by the geometrical factor which leads in the final analysis to the additive dependence of the molar volumes.

The surface tension polytherms are complicated. With an increase of the bismuth content, the temperature coefficient of surface tension decreases in the absolute value. In the alloy with 0.03 molar fractions of bismuth, examination showed the phenomenon of temperature buffering of surface tension. A further increase of the bismuth content results in a non-linear dependence of the surface tension on temperature; these relationships of the temperature dependence of surface tension of solutions at strong adsorption of one of the components have been detected previously by other authors [176, 214] and are explained by the desorption of the active component by the surface with increasing temperature.

The surface tension isotherms are most interesting (Fig. 4.15). Bismuth greatly reduces the surface tension of gallium even at low concentrations indicating its high surface activity. A further increase of the bismuth content results in a monotonic decrease of the surface tension to the value for pure bismuth. As indicated by Fig. 4.15, the surface tension isotherms in the gallium–bismuth systems are characterised by smooth curves without singular points.

The calculations of surface tension using the currently available theoretical equations require, in addition to others, the data on the concentration dependence of the partial molar volumes and surfaces. In the investigated system, the molar volumes are governed by the additive law, but the molar volumes of the pure components differ by more than two orders of magnitude. As shown in section 2.5, this should should result in large errors in the determination of the partial–molar components, using the given approximate equation. Therefore, calculations of the partial–molar surface of the components in the discussed systems have been carried out using the equations (2.6.7) and (2.6.8). The results show that the maximum differences are found in the concentration ranges enriched by one of the components. For example, for the partial–molar surfaces of gallium differences greater than 5% were found at 0.98 molar fractions of bismuth.

Using these results, we calculated the surface tension of gallium–bismuth melts using the well-known Popel'–Pavlov equation. The results of the calculation show that the differences in the determination of the partial molar surfaces have almost no effect on the calculated value of the surface tension and this confirms the conclusions made in the second chapter.

Figure 4.15 shows the experimental and theoretical isotherms of surface tension. The graph shows that the Popel'–Pavlov equation efficiently describes the concentration dependence of surface tension in the gallium–bismuth system.

The experimental data were used to calculate: adsorption (N-variant), average composition and the effective thickness of the surface layer.

The results of the calculations show that in the melts diluted with respect to bismuth the adsorption of bismuth rapidly increases.

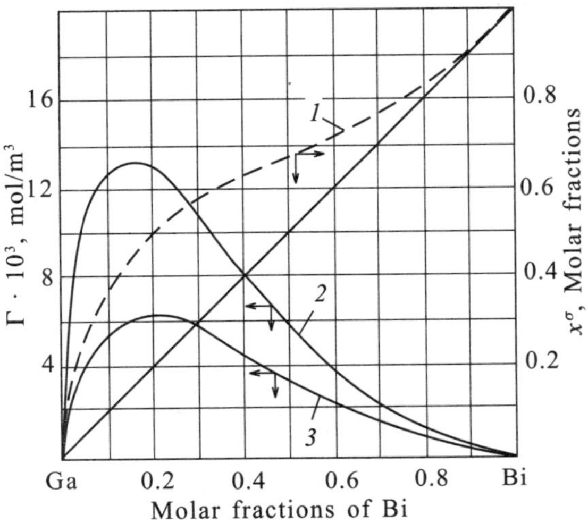

Figure 4.16. Isotherms of adsorption and molar fractions of bismuth in the surface layer of gallium–bismuth melts. 1) the content of bismuth in the surface layer at 773 K; 2, 3) the isotherms of adsorption of bismuth (N – variant according to Guggenheim and Adam) at 573 and 773 K.

The isotherms reaches the maximum value at 0.2 molar fractions of bismuth. A further increase of the bismuth content results in a monotonic decrease of the isotherms to zero. It is interesting to note that in the alloys with the bismuth content of 0–0.4 molar fractions the desorption of bismuth is intensive and increases with increasing temperature. These alloys are also characterised by the non-linear dependence of surface tension on temperature and this confirms the explanation proposed previously for this phenomenon.

The isotherms of the molar fractions of bismuth in the surface layer, presented in Fig. 4.16, indicate that the surface layer of all the gallium–bismuth alloys is enriched with bismuth atoms. The largest difference between the composition of the surface layer averaged over the entire thickness in relation to the composition of the solution is detected in the alloys with the molar fractions of bismuth of 0.2.

Thus, the experimental investigations of the gallium–bismuth melts indicate the absence of the previously observed special features on the surface tension isotherms of this system, and the resultant isotherms are efficiently described by the Popel'–Pavlov equation.

5

Surface tension and density of multicomponent melts

Regardless of the extensive application of a large number of component metallic systems, in particular melts [163, 164], in technology, the surface properties of these systems have been studied insufficiently. At the moment of definition and formulation of the tasks of the work described in this book, the surface tension of only nine ternary metallic systems had been studied. As far as we know, the dependence of σ on the composition of melts with a large number of components has not been studied.

This state is determined by the large increase of the number of experimental measurements taken in the examination of the properties of multicomponent melts. On the other hand, the increase of the number of components greatly complicates both the theoretical description of the variation of surface tension with composition and temperature and also the calculation, using the experimental data, of the main parameters of the surface layer.

It is evident that the absence of a reliable experimental base has a negative effect on the efficiency of solving applied problems and inhibits the development of the theory of surface phenomena in the multicomponent systems. Therefore, we have examined the surface tension and density of seven ternary (In–Sn–Ga, In–Sn–Pb, Th–Pb–Bi, In–Sn–Tl, Tl–Pb–In, Tl–Pb–Sn, In–Pb–Bi) and two four-component (In–Sn–Pb–Bi, In–Pb–Bi–Ga) systems. The selection of the systems was dictated by the following considerations: extensive use in practice [222]; comparatively low melting point; the absence of interaction with glass from which the device is produced. For further clarification and development of theory, it is interesting to examine the effect of the third and fourth component on the special features of the surface tension isotherm of binary systems.

In the investigated melts, complicated forms of the isotherms of σ are typical of binary systems: In–Sn, Tl–Pb, Ga–Pb. Consequently, it should be assumed that the concentration dependence of σ in these multicomponent systems is also complicated.

The In–Sn–Pb, In–Sn–Ga, Tl–Pb–Bi systems and the In–Sn–Pb–Bi four-component system were investigated in the combined device described in section 3.5.

The temperature dependence of the surface tension and density of 60–100 alloys was measured in each system. Consequently, it was possible to determine, with a high degree of reliability, the special features of the temperature and concentration dependence of these properties.

The application of the method of mathematical experiment design [223] enabled the number of experiments to be greatly reduced. However, this increased the requirements on the accuracy and availability of the experimental data. These requirements are satisfied by the device described in the section 3.2.2.1. Therefore, the In–Sn–Tl, In–Pb–Tl, Sn–Pb–Tl and In–Pb–Bi ternary systems were investigated in this device.

The In–Pb–Bi–Ga melts undergo phase separation at temperatures below 923 K. Therefore, the system was investigated using the methods of experiment design in the equipment described in detail in [215, 239]. In this equipment, the surface tension and density were measured by the sessile droplet method from the critical phase separation temperature to 1123 K.

5.1. Surface tension and molar volumes of ternary systems

5.1.1. The indium–tin–lead system

In this system, the region of composition adjacent to the indium–tin binary system has been studied in greater detail. Consequently, for a relatively small depth of the minimum on the surface tension isotherms of the indium–tin binary system, it was possible to determine the effect of surface-active tin on the system. To construct the isothermal surfaces of the physical–chemical properties, measurements were taken of the temperature dependence of surface tension and density of more than 100 alloys with the composition distributed both on the sections of the concentration angle (x_{In}:x_{Sn}=1:4,1:1, 3:1, 9:1; x_{Sn}:x_{Pb}=13:1, 3:1; x_{In}:x_{Pb}=1:5, 2:3) and in the section with a constant lead content (x_{Pb} = 0.0335 molar fractions).

The results of the measurements show that the temperature dependence of the density of all investigated alloys is linear with the negative temperature coefficient. The values of the coefficient of linear approximation and of the ranges of the coefficients at a fiducial probability of 0.95 published in Table 11.

The experimental data were used to calculate the molar volumes of ternary melts at different temperatures. The isotherms of the molar volumes in indium–lead [192, 225] and tin–lead lateral binary systems [226, 227] are characterised by small positive deviations from additivity. The positive deviations were also find in our work on indium–tin melts. The isotherms of the molar volumes in the beam sections of the ternary system are shown in Fig. 5.1. They also have positive deviations from the additivity.

With the increase of the lead content of the melts, the values of the molar volumes approached the additive values. The deviations of the isotherms of the molar volumes from additive straight lines

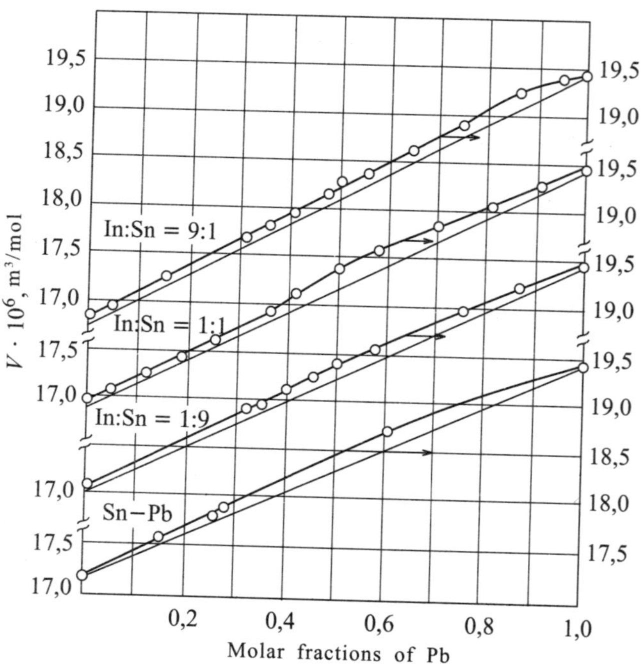

Figure 5.1. The isotherms of molar volumes in the indium–tin–lead system at 623 K. 1) the isotherms of surface tension at 723 K; 2,3) the lead content in the surface layer of the melts at the 573 and 723 K, respectively; 4) the indium content in the surface layer of the melts at 723 K.

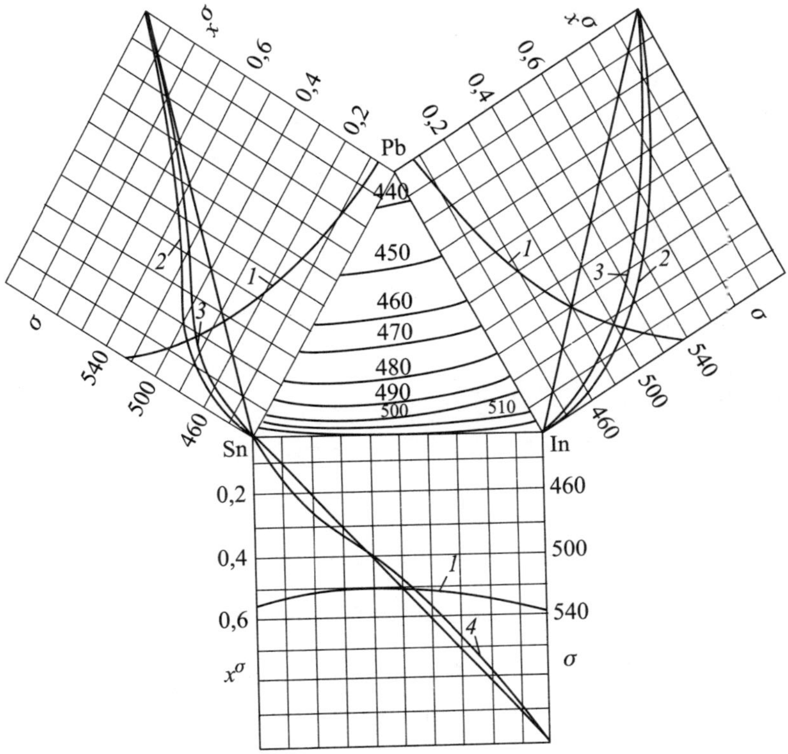

Figure 5.2. The isolines of surface tension (mJ/m²) in the indium–tin–lead system at 723 K. 1) the indium–lead system; 2) section with $x_{In}: x_{Sn}$ = 1:1; 3) the tin–lead system.

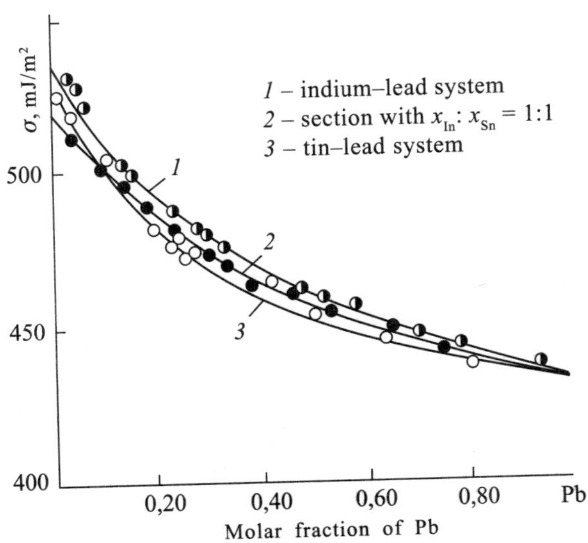

Figure 5.3. The isotherms of surface tension at 773 K.

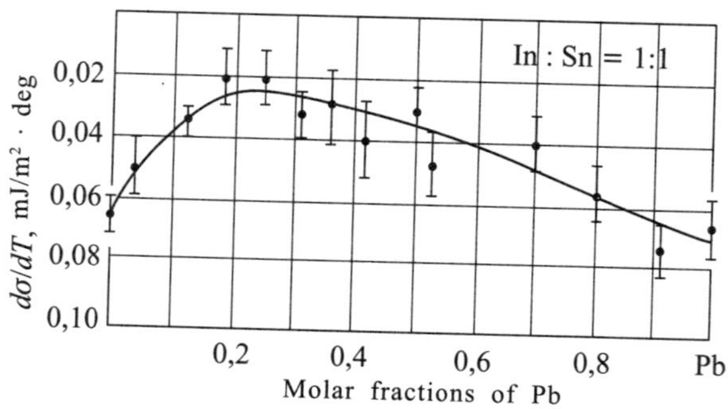

Figure 5.4. The dependence of the temperature coefficient of surface tension on the lead content in the indium–tin–lead melts.

in ternary melts do not exceed the identical deviations in lateral binary systems.

The temperature dependence of the surface tension of the investigated alloys is linear with a negative temperature coefficient. The variation of the temperature coefficient of surface tension $\left(\frac{d\sigma}{dT}\right)$ with the composition is shown in Fig. 5.4 which indicates that with the increase of the lead content the absolute value of $\left(\frac{d\sigma}{dT}\right)$ decreases, approaching zero at x_{Pb}=0.25. A further increase of the lead content results in a linear increase of the absolute value of the temperature coefficient. It is interesting to note that the identical dependence of $\left(\frac{d\sigma}{dT}\right)$ on the composition of the alloy was found in indium–lead [228] and tin–lead [219] binary systems, and in all sections with a constant ratio of the molar fractions of indium to tin.

This nature of the variation of the temperature coefficient of surface tension may be easily understood by examining the adsorption phenomena. The results of calculations of adsorption in the ternary system show (section 5.1.1) that in the diluted solutions with up to 0.2 molar fractions of lead there is a considerable positive adsorption of lead. Increase of temperature results in a change of surface tension. Two opposite factors act in this case. On the one hand, with the increase of temperature the atomic spacing increases resulting in a weakening of the intermolecular forces and, consequently, in a decrease of surface tension. However, a decrease of the surface tension is prevented by desorption of lead with the increase of temperature. The value of the temperature coefficient of surface tension depends on the ratio of these factors. If these factors in the alloy with 0.2 molar fractions of lead almost completely cancel each other out, then with the increase of the lead concentration the role of desorption becomes insignificant.

Thus, the isothermal surfaces of the molar volumes and surface tension of the indium–tin–lead ternary system are described by the smooth surfaces without any singular regions. The isotherms of the investigated properties $\left(\rho,\sigma,\frac{d\sigma}{dT}\right)$ in the sections of the concentration angle do not contain any special features not predicted by the isotherms of these properties in the lateral binary systems.

5.1.2. The thallium–lead–bismuth system

As shown previously, the isotherms of the surface tension of the thallium–lead binary system, like in In–Sn, are characterised by flat minima in the region of medium compositions. Like lead in alloys with indium and tin, bismuth is a surface-active component in the alloys with lead and thallium. It should be expected that the relationships governing the variation of the surface tension with composition in this system will be of the same nature as in the indium–tin–lead ternary melts.

In the thallium–lead–bismuth systems, we examined the temperature dependence of the density and surface tension of the alloys whose compositions are situated on the sections of the concentration angle with a constant ratio of the molar fractions of thallium and lead (1:9, 1:3, 1:1, 3:1, 9:1) (Table 13).

Mathematical processing of the experimental data for the densities of the melts by the method of least squares shows that the errors of determination have positive deviations from additivity. As indicated by Fig. 5.5, the largest deviations are found in the thallium–bismuth melt and in the alloys adjacent to this system. When replacing thallium by lead in the cross-section of the concentration angle, the isotherms of the molar volumes approached the additive straight lines and in the binary lead–bismuth system they are almost identical with these lines [230, 231].

With increasing temperature, the isotherms of the molar volumes approach the additive straight lines. The temperature dependence of the surface tension of all alloys of the ternary system is linear with a negative temperature coefficient. The results of linear

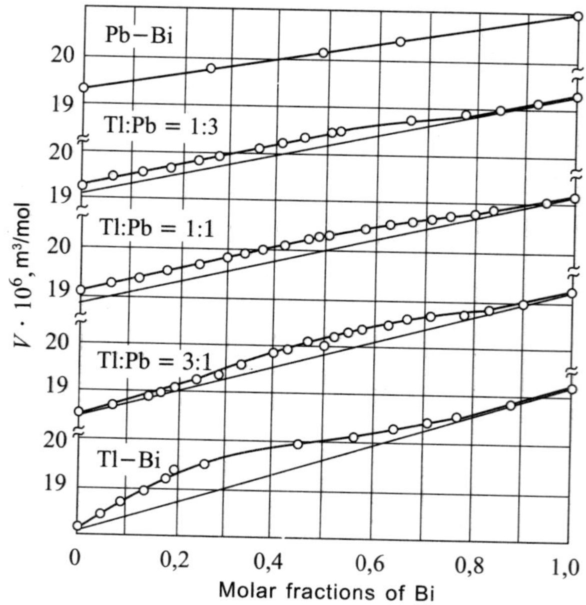

Figure. 5.5. Isotherms of molar volumes (m^3/mol) in the thallium–lead–bismuth system at 623 K.

approximation of the resultant data are presented in Table 14. The ranges of the coefficient of the approximation equation are given for 95% probability. No special features were found on the concentration dependence of the temperature coefficient of surface tension. The values of $\frac{d\sigma}{dT}$ for the alloys of the ternary system do not exceed the values for the pure metals.

The isotherms of surface tension for the investigated sections of the concentration angle are characterised by smooth curves without singular points. With increase of the bismuth content of the alloy, the surface tension monotonically decreases. Bismuth is surface-active both on thallium and on lead, and also in all thallium–lead alloys (Fig. 5.6). However, its limiting surface activity on thallium is considerably higher than on lead and increases in a linear manner for the alloys when lead is replaced by thallium.

As in the indium–tin–lead system, the isothermal surface of σ of the investigated system is characterised by the presence of a 'depression' (Fig. 5.7) which forms a flat minimum on the isotherms of surface tension of the thallium–lead binary system and sections with a constant bismuth content. With the increase of the bismuth content in the section, the depth of the minimum on the isotherms

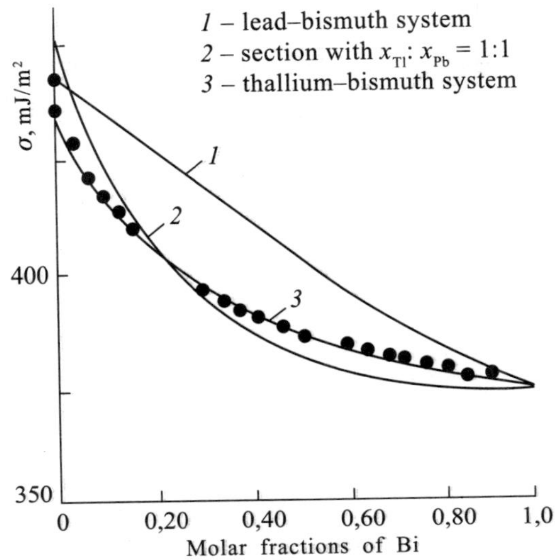

Figure 5.6. The isotherms of surface tension (mJ/m²) in the thallium–lead–bismuth system. 1) the lead–bismuth system; 2) the section x_{Tl}:x_{Pb} = 1:1; 3) the thallium–bismuth system at 773 K.

of σ decreases. The position of the minimum is displaced towards the Tl–Bi side of the triangle. This is caused by the different limiting surface activity of bismuth on pure metals such as thallium and lead and on alloys of the thallium–lead binary system. As indicated by Fig. 5.6, the isotherms of σ of the thallium–bismuth and lead–bismuth binary systems and of the ternary section intersect. Since these curves intersect at different bismuth concentrations, this does not cause any concentration buffering of surface tension.

The relationships governing the limiting surface activity of lead and thallium in the thallium–lead–bismuth ternary system are identical. Lead shows surface activity on thallium and is inactive on bismuth. With the increase of the bismuth content in the alloys with thallium, the value of the limiting surface activity of lead rapidly decreases, passing through zero at x_{Bi}= 0.0 fractions of bismuth. The identical relationship is characteristic of the limiting surface activity of thallium in the lead–bismuth alloys.

5.1.3. The indium–tin–gallium system

A common feature of the isothermal surface of the surface tension of the previously examined ternary alloys (In–Sn–Pb and thallium–Pb–Bi) is the presence of minima on the isotherms of σ of one of

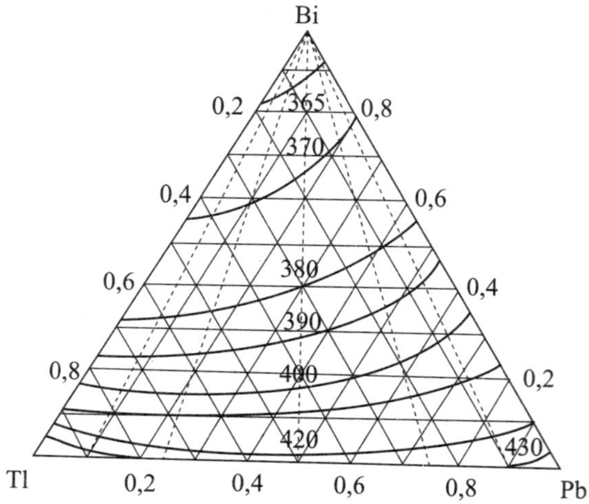

Figure 5.7. Isolines of surface tension of the thallium–lead–bismuth system.

the lateral binary systems. The third component in both systems is represented by surface-active lead (system In–Sn–Pb) and bismuth (system Tl–Pb–Bi). It is interesting to examine the effect of the inactive component on the special features of the isotherms of σ of the binary systems. For this purpose, the indium–tin–gallium system appears to be highly suitable. In this system, gallium is inactive both in indium and in tin. On the other hand, the isotherms of σ of the tin–gallium binary system, according to [218], are characterised by complicated curves with a minimum and a maximum. Consequently, the isothermal surfaces of σ of the indium–tin–gallium ternary system may also prove to be complicated.

The temperature dependence of the surface tension and density was investigated on the alloys of the ternary system whose composition is situated on the sections with the constant ratio of the molar fractions of indium to tin ($x_{In}:x_{Sn}$ = 1:9, 1:3, 1:1, 9:1). In addition to these data, the isothermal surface was constructed using experimental data determined for pure metals and lateral binary systems.

The temperature dependence of the density of all examined alloys is linear. Table 15 in the Appendix gives the values of the coefficients of the linear equation and their confidence ranges.

The molar volumes of the investigated ternary alloys within the limits of the determination error coincide with the additive values. As an example, Fig. 5.8 shows the isotherms of molar volumes with the variation of the composition in the section with the constant ratio

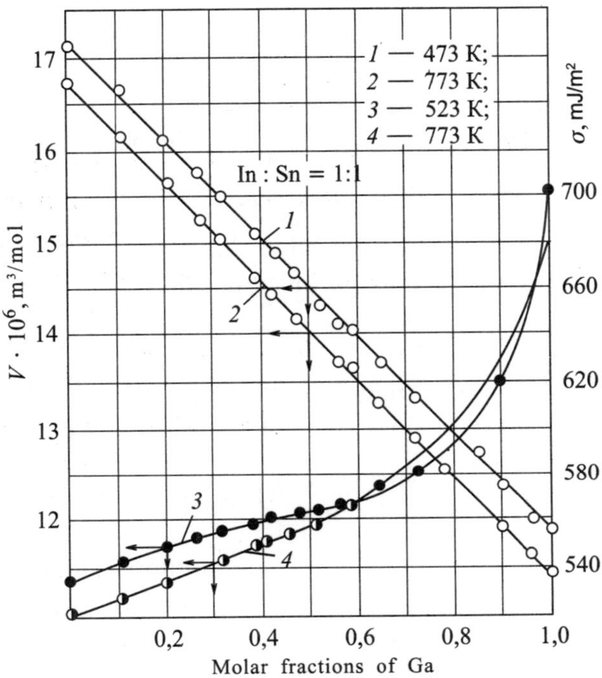

Figure 5.8. Isotherms of surface tension (473 K, 773 K) and molar volumes (523 K, 773 K) in the indium–tin–gallium system. 1) 473 K; 2) 773 K; 3) 523 K; 4) 773 K.

of the molar fractions $x_{In}:x_{Sn}=1:1$.

The additive isotherms of the molar volumes were also obtained for other sections of the ternary system, indicating the additivity of the isothermal surface of the molar volumes in the indium–tin–gallium ternary system. This conclusion is confirmed by the experimental data for lateral binary systems: tin–gallium [218], indium–gallium [217, 190], indium–tin [211, 232] and are in agreement with the results of the measurements of the heats of mixing [234] and the equilibrium diagram of the ternary system [235].

It should be mentioned that in [233] investigations are carried out on 13 alloys of this ternary system and the results showed large deviations of the molar volumes from the additivity in the range of the medium compositions. Our data do not confirm this result.

Thus, the experimental results show that the isothermal surface of the molar volumes in the indium–tin–gallium ternary system coincides with the additive plane.

Figure 5.9 shows the polytherms of surface tension of the ternary melts, indicating that the temperature dependence of σ is far from

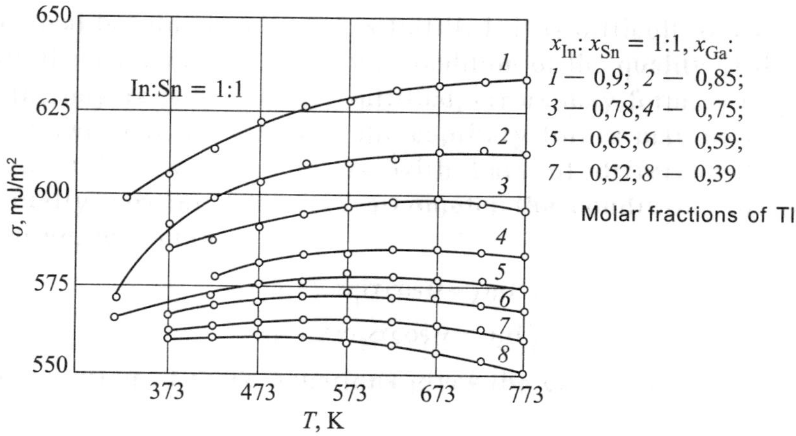

Figure 5.9. Polytherms of the surface tension of indium–tin–gallium melts.

linear. The non-linear form of the polytherms of σ is evident in the entire investigated temperature range. The non-linear dependence of surface tension σ on temperature was detected both in the indium–gallium and lead–gallium binary systems and in ternary indium–tin–gallium systems at a gallium content higher than 0.25 molar fractions. The detected special features of the polytherms may be explained, as shown in [13], by the simultaneous variation of the quantities with temperature: partial tensions (the work function of the components, related to the unit surface), partial–molar surfaces, and adsorption of the components. If the first two factors determine a decrease of surface tension, the third factor increases the surface tension. Here, the fraction of each factor depends on the special features of the system and temperature leading to a change of the magnitude nd sign of the temperature coefficient and to the non–linear form of the polytherms of σ. This problem was examined in considerable detail in [13].

The results of measurements of the surface tension of a melt of the indium–tin–gallium ternary system are presented in Table 16.

In addition to the isotherms of the molar volumes, Fig. 5.8 shows the concentration dependences of surface tension with the variation of composition in the beam sections of the concentration triangle (x_{In}:x_{Sn}=1:1). The graph shows that the isotherms of σ at 523 K and 773 K are described by the smooth curves, bent in the direction of the composition axis. At 527 K, the isotherms have an inflection point at 0.6 molar fractions of gallium. With increasing temperature

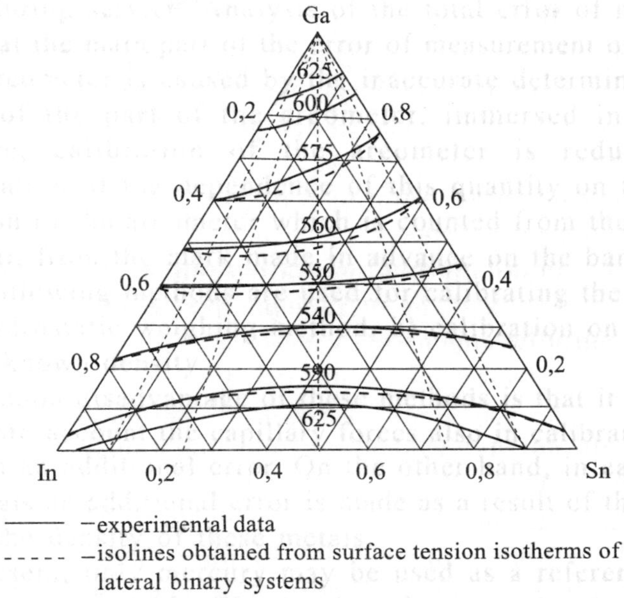

Figure. 5.10. Surface tension isolines at 773 K in the indium–tin–gallium system.

of the inflection point becomes smoother, as illustrated by the isotherm at 773 K. it should be mentioned that the isotherms σ at different temperatures intersect at two points (at 0.6 and 0.98 molar fractions of gallium). These alloys are characterised by the well-known phenomenon of temperature buffering of surface tension, *i.e.* these alloys are characterised by the fact that the surface tension is independent of temperature.

Gallium increases the surface tension of both pure metals like indium and tin, and also of all the alloys of the indium–tin binary system. With the variation of the composition along the lines with a constant gallium content, the surface tension isotherms coincide up to 0.4 molar fractions of gallium with the isotherms of the indium–tin binary system, *i.e.*, a flat maximum appears on the isotherms. With a further increase of the gallium content, the position of the minimum is displaced in the direction of the Ga–Sn side and the depth of the minimum decreases.

Figure 5.10 shows surface tension isolines providing more complete information on the isothermal surface σ. As in the indium–tin–lead and thallium–lead–bismuth systems, the isothermal surface σ of the indium–tin–gallium systems shows a depression forming minima on the isotherms σ of the indium–tin binary system and ternary sections.

However, if in the In–Sn–Pb and Tl–Pb–Bi ternary systems the depression is smoothed out at 0.08 and 0.20 molar fractions of lead and bismuth, respectively, in the investigated system, the 'depression' is detected up to 0.4 molar fractions of gallium. The reasons for this behaviour are discussed in chapter 6 on the basis of analysis of adsorption processes in these smells.

5.2. Application of the methods of mathematical experimental design in examination of the surface properties of multicomponent systems

As indicated by the above considerations, experimental examination of the properties of the multicomponent metallic melts requires a large amount of time and highly qualified labour. In particular, in the indium–tin–lead ternary systems we have examined approximately 100 alloys of different composition. Simple calculations show that more than 25 000 measurements have been made in this case. With the increase of the number of components, the volume of work increases in accordance with a geometrical progression. Evidently, these methods cannot be used to examine the properties of even the multicomponent melts which are used in practice.

The number of experiments can be greatly reduced, without losing any important information, by the methods of mathematical experimental design [223]. Special attention has been paid to the development of these methods in recent years [236–240].

The investigations of the concentration dependence of the physical–chemical properties of the multicomponent systems are usually carried out using the method of simplex lattices [237, 239–241]. The method is based on the construction of a mathematical model linking of the investigated property with the composition of the solution. In this case, the factorial space is represented by a regular simplex with g-tips in (g–1)-measurement. Each point of the simplex corresponds to the solution of a specific composition and, vice versa, a specific simplex point corresponds to the solution of any composition.

In a general case, the 'composition–property' mathematical model is constructed using the canonical form of the polynomial of degree n [237, 240]:

$$y=\sum_{i=1}^{g}\beta_i x_i+\sum_{m=2}^{n}\left\{\sum_{i,j=1}^{g}\beta_{ij}^{m}x_i x_j\left(x_i-x_j\right)^{m-2}\right\}+\sum_{m=3}^{n}\left\{\sum_{1\le i\le i_2\ldots i_m\le g}^{g}\beta_s x_{i1}^{s_1}x_{i2}^{s_2}\ldots x_{im}^{s_m}\right\} \quad (5.2.1)$$

The investigated property is presented in the form of the polynomial (5.1). To determine the coefficients of the given model, it is necessary to select specific points of the experimental plan and determine the value of the investigated property.

In [237] plans were proposed for these purposes ensuring the uniform scatter of the experimental points over the $(g-1)$-dimensional simplex. The points of these plans are represented by the nodes of the simplex lattices whose number is equal to the number of coefficients in the polynomial.

In a general case, the coefficients of the polynomial (5.2.1) are determined by a system of equations derived by substituting the values of the function y_{ijk} and the coordinates of all points of the plan x_{ijk}.

The formulae for calculating the coefficient are greatly simplified using the reduced polynomial.

The models of different orders, used in examination of the multicomponent systems, and also the expressions for the polynomial coefficient, are presented in [223].

Calculating the polynomial coefficients from these formulae gives the approximating model for predicting the value of the investigated property at any point of the simplex.

To determine the accuracy of prediction of the response function, given in the form of the polynomial (5.2.1), the adequacy of the latter is evaluated. It is assumed that the values obtained for every point are distributed with a normal dispersion [223].

To verify adequacy, several additional verification experiments must be carried out. Usually, they are selected in complicated sections in which it is assumed that the deviations of the predicted values of the functions from the true values are maximum. If the resultant equations describe insufficiently the true dependence of the physical–chemical property on the composition, it is necessary to transfer to a model of a higher order and carry out further examination of the functions in singular regions.

In conclusion, it should be mentioned that the application of the method of experiment design is also associated with certain difficulties. A shortage of prior information often requires a large number of experiments to be carried out and this cast doubts on the efficiency of application of cumbersome mathematical methods. Therefore, the improvement of the existing methods of design with special reference to investigation of the specific properties is of considerable importance. These methods were used for the first time for determination of surface tension in [239]. It was shown that the optimal degree of the polynomial is the one which describes with sufficient accuracy the concentration dependence of the surface tension of lateral binary systems.

On the basis of this conclusion it is possible to select the optimal model for describing the multicomponent system using the data on the properties of lateral binary solutions.

5.3. Investigation of the surface tension and density of ternary systems using experiment design methods

This section is concerned with the results of experimental examination of the surface tension and density of ternary systems: In–Pb–Tl, In–Sn–Tl, Sn–Pb–Tl and In–Pb–Bi.

Investigations were carried out using the simplex-lattice plans. This resulted in a large decrease of the number of experiments, whilst retaining the reliability of the results. The method of determination of the order of the coefficients of the regression equations was described previously.

The polynomials of different degrees and equations for calculating the coefficients were taken from [223].

Measurements of surface tension and density were taken in a combined device, whose design features are described in section 3, 4.1. Main advantages of the device are the high accuracy of the measurements and the possibility of preparing an alloy with a strictly defined composition which becomes controlling in experimental design.

Analysis of the isotherms of the surface tensions of natural binary systems shows that to describe the concentration dependence of the discussed ternary system, the model of the fourth order appears to be optimum. To determine the polynomial coefficients and also to verify the adequacy of the selected model, alloys of strictly given compositions were investigated in each ternary system. The resultant experimental data on the temperature dependence of surface tension and density are presented in Tables 17–20 of the Appendix.

5.3.1. Concentration dependence of the surface tension of ternary systems

The results of experimental investigations show that the dependence of density on the composition in indium–tin–gallium, tin–lead–thallium and indium–lead–thallium ternary systems in the experimental error range is described by an additive function.

In the indium–bismuth–thallium melts, the dependence of density on composition is also close to additive. Only the indium–bismuth binary system and ternary alloys, adjacent to this side of the composition angle, showed small positive deviations of density from additive values.

It is well known that density is a volume-additive physical–chemical property. Therefore, the result which can be used for efficient and rapid determination of the density of the ternary alloy of any composition, is of considerable importance for practice.

The isothermal surfaces of the molar volumes are characterised by mainly positive deviations from the additive plane. However, in the investigated ternary alloys, they do not exceed the deviations of the molar volumes from the additive values in the lateral binary systems.

According to the literature data, the surface tension isotherms of the In–Sn, Sn–Tl, Sn–Tl and Pb–Tl systems, are characterised by a minimum in the region of medium compositions. Therefore, in addition to the ternary melts, we have examined these systems in detail. As shown in chapter 4, the literature data on the In–Sn and Sn–Tl binary melt have been confirmed. However, the experimental data on the In–Tl and Sn–Tl melds indicate the absence of the special features on the isotherms of σ of these systems detected previously by other authors.

Comparison of the Figs. 4.8, 4.9 and 5.11 shows that the surface tension isotherms of the sections of the ternary systems are close

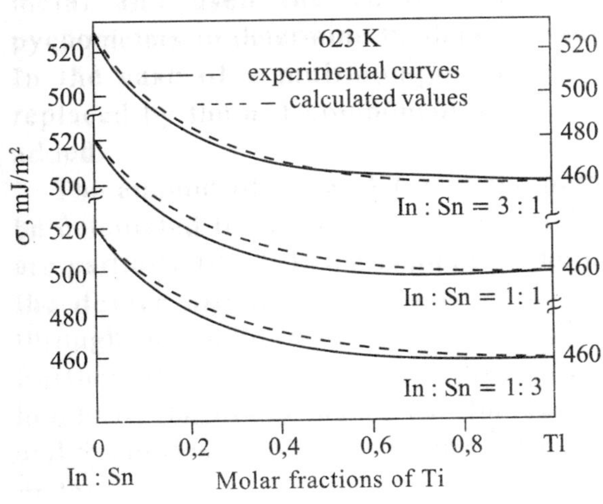

Figure 5.11. Surface tension isotherms of indium–tin–gallium melts at 623 K; solid lines the experimental data, dashed lines the calculated values.

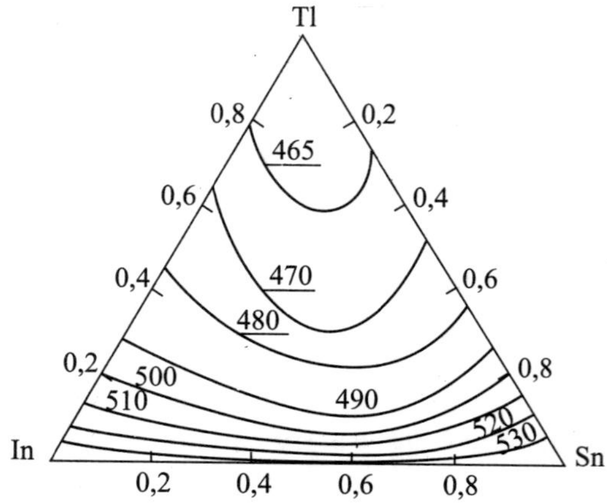

Figure 5.12. Surface tension isolines (mJ/m²) at 623 K in the indium–tin–thallium system.

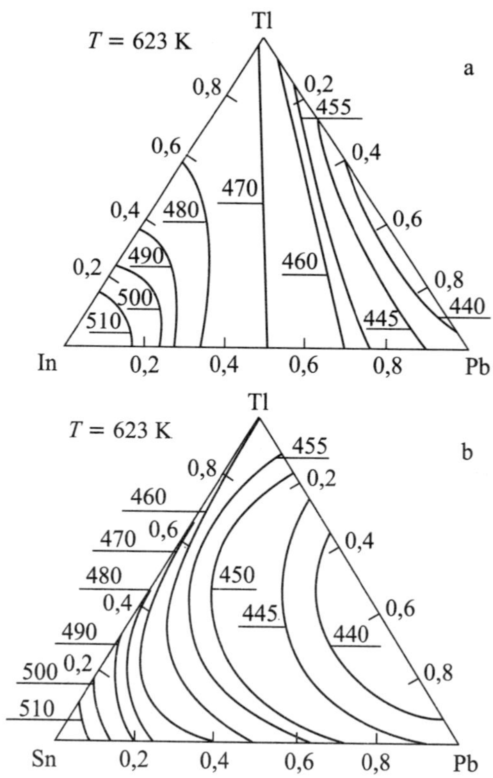

Figure 5.13. Surface tension isolines (mJ/m²) at 623 K in indium–lead–thallium and tin–lead–thallium systems.

in the form to the isotherms of σ of the indium–thallium and tin–thallium binary systems and have the form of smooth curves without any singularities.

The concentration dependences of σ in the investigated ternary systems have many common features. However, the isothermal surfaces of σ of every ternary system have their own special features which will be discussed separately.

a. The indium–tin–gallium system

As mentioned previously, the isothermal surfaces of σ are described by the polynomial of the fourth degree. To provide more information, Figs. 5.14 and 5.15 give both the surface tension isolines and the isotherms of σ for different sections of the composition angle of the indium–tin–gallium system. Figure 5.14 shows that thallium reduces the surface tension of both indium and tin pure metals and of all alloys of the In–Sn binary system. The limiting surface activity of thallium on indium is higher than on tin. With replacement of indium in the alloys with tin, the limiting surface activity of thallium in the indium–tin alloys changes only slightly.

The graph also shows that the isotherms of σ in the beam sections of the ternary system are situated below the isotherms of the indium–thallium and tin–thallium binary systems. This indicates the presence of a flat minimum on the isotherms of σ with respect to the dimensions corresponding to the constant thallium content.

The surface tension isolines (Fig. 5.15) provide more complete information on the relief of the isothermal surface of the property. The surface of σ of the In–Sn–Tl ternary system is characterised by a 'depression', forming a minimum on the isotherms of the In–Sn binary system and the sections of the ternary system with a constant gallium content. With the increase of the thallium content, the depth of the 'depression' decreases. However, in contrast to the In–Sn–Pb and In–Sn–Ga melts, the 'depression' on the isothermal surface of σ of the investigated system is retained up to pure gallium. This indicates that the presence of the atoms of the surface-active gallium has no significant effect on the special features of the concentration dependence of surface tension in the indium–tin binary system, at least it does not change quantitatively the form of isotherms of σ.

b. The tin–lead–thallium system

The concentration dependence of surface tension in the tin–lead–thallium ternary system is similar in its form to the dependence in the indium–tin–thallium melts. As indicated by Fig. 5.13, the isothermal surface σ of this system is also characterised by a 'depression' forming minima on the isotherms of σ of the lead–thallium melts and sections with a constant tin content. However, in contrast to the indium–tin–thallium system, in the investigated case the increase of the tin content is accompanied by the displacement of the minimum on the surface tension isotherms of the ternary sections in the range of the mean composition in the direction of the Tl–Sn concentration angle. The depth of the minimum decreases. At 0.7 molar fractions of tin, the 'depression' disappears.

Figure 5.14 shows the isotherms of surface tension in the beam sections of the concentration angle. The graph indicates that the introduction of thallium is accompanied by a decrease of the value of σ of both tin and lead and also of all tin–lead alloys. However, the limiting surface activity of thallium in tin is higher than in lead and increases in a linear manner when tin is replaced with lead.

Attention should also be given to the fact that the minima on the surface tension isotherms are also detected on the beam sections of the concentration angle with the constant ratio of the molar fractions of tin and lead. In this case, the minimum is retained up to the section with the ratio $x_{Sn}: x_{Pb}$=3:1. This is caused by the fact

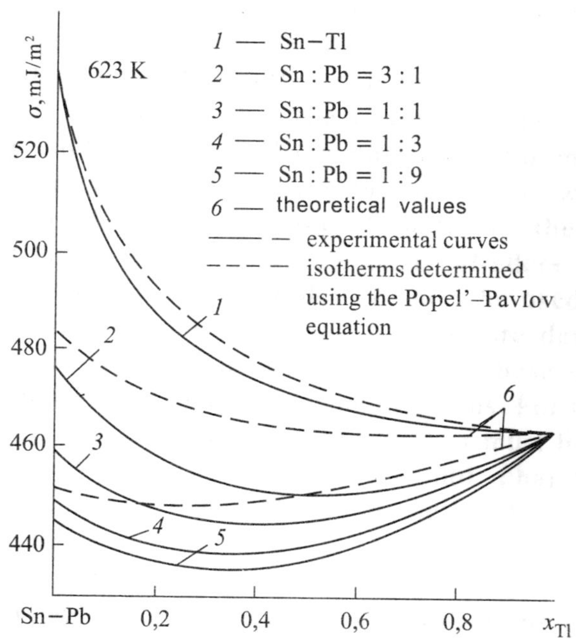

Figure 5.14. The surface tension isotherms at 623 K in the tin–lead–thallium system.

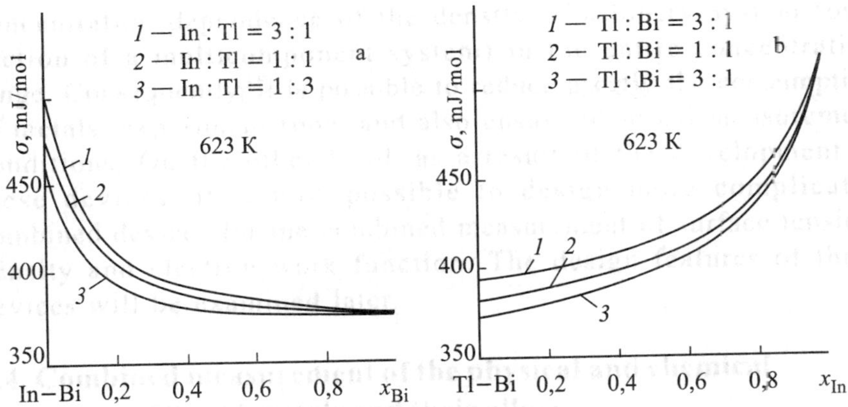

Figure 5.15. The surface tension isotherms (mJ/m²) at 623 K in the indium–thallium–dismal system.

that the tin atoms (tin is an inactive component) concentrate mainly in the volume of the melt and have no significant effect on the concentration dependence of surface tension. Thus, in a wide range of the compositions, with the exception of a small section adjacent to the Sn tip, the form of the isothermal surface of σ is determined by the surface-active components thallium and lead. Consequently, the processes taking place in the surface layer are determined mainly by the ratio of the atoms of these components. Evidently, this relationship is general and characteristic of multicomponent systems.

c. The indium–bismuth–thallium system

The surface tension isotherms in the beam sections of the concentration angle of the In–Bi–Tl ternary system are shown in Fig. 5.15. As indicated by the graphs, the isotherms of σ are smooth curves without any extreme points.

Bismuth is a surface-active component on both pure metals gallium and indium and in In–Tl alloys. Although the difference between the values of surface tension of indium and bismuth is larger than between thallium and bismuth, the limiting surface activity of bismuth in these metals is the same.

Indium is inactive in both pure metals and in binary alloys, and also in all alloys of the ternary system.

A characteristic feature of the effect of thallium is the change of the sign of surface activity with the increase of the bismuth content. From the surface active element, bismuth becomes inactive.

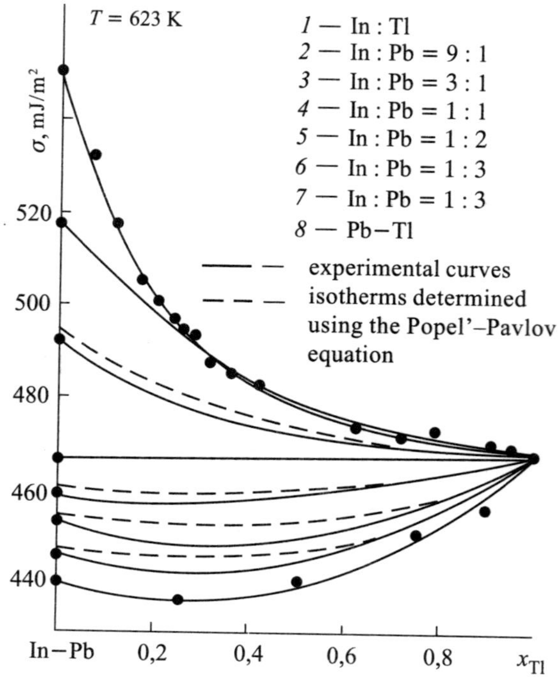

Figure 5.16. The surface tension isotherms at 623 K in the indium–lead–thallium ternary system.

This phenomenon, typical of many ternary systems, could be regarded as the phenomenon of concentration inversion of surface tension.

Thus, the isothermal surface of the indium–bismuth–thallium ternary system is smooth, without any special regions. The surface tension isotherms in the quasi-binary sections of the ternary system do not contain any qualitatively new special features in comparison with the isotherms of the lateral binary systems.

d. The indium–lead–gallium system

As indicated by Figs. 5.13 and 5.16, the isothermal surface of σ of the indium–lead–thallium ternary system has many common features with the tin–lead–thallium system. Lead increases the surface tension of indium, thallium and all indium–thallium melts. The behaviour of thallium in the lead–indium systems is identical. The isotherms of σ of both the thallium–lead binary system and of the sections of the concentration triangle adjacent to the Tl–Pb side are characterised by flat minima in the range of medium compositions.

However, in contrast to the Sn–Pb–T1 melts, in the discussed system the minimum on the isotherms of σ of the ternary sections disappears at a relatively low content of indium (0.25 molar fractions). On the isothermal surface of σ the special feature is reflected in the form of a localised 'depression' in the small region in the immediate vicinity of the Tl–Sn side. The typical feature of the concentration dependence of σ in the investigated system is the presence of a buffer section, *i.e.*, the section of the concentration triangle with a constant value of surface tension. In fact, Figs. 5.13 and 5.16 show that with the variation of the composition in the section with the constant ratio of the molar fractions of indium to lead $x_{In}:x_{Pb} = 1$, the surface tension is independent of the thallium content, *i.e.*

$$\left(\frac{\partial\sigma}{\partial x_{Tl}}\right)_{X_{In}:X_{Pb}} = 0$$

This phenomenon in water and salt solutions was detected a long time ago and referred to as the concentration buffering of surface tension. Theoretical analysis and a review of the currently available experimental data on concentration buffering of σ has been published in [6, 13, 19, 251]. However, analysis of these studies shows that the literature data are ambiguous even in the definition of the phenomenon of concentration buffering of surface tension. This is caused by the existence of a large variety of the possible mechanisms of variation of the composition in the multicomponent systems, in contrast to the binary systems in which only a single mechanism operates ($dx_1 = -dx_2$). Therefore, this problem is examined in detail in chapter 5 for a general case of multicomponent systems.

5.4. Surface tension of four-component systems

In this section, we discuss the results of experimental examination of the surface tension and density of indium–tin–lead–bismuth and indium–lead–bismuth–thallium four-component systems [242].

The In–Sn–Pb–Bi alloys were investigated in the same combined device (3.5) as the one used for measuring the surface tension by the method of maximum pressure in the droplets, and the density was investigated using an improved areometer. The melts of the second quaternary system (In–Bi–Pb–Ga) showed phase separation at a temperature below 873 K. Therefore, measurements of the surface tension and density of these alloys were carried out by R.U. Goitemirov together with the author of the experimental equipment described in detail in [215, 239].

These systems were selected for investigations because the isotherms of σ of the In–Sn, Ga–Bi and Ga–Pb lateral binary systems are complicated, as indicated by the literature data [127, 248–250]. Therefore, the dependences of surface tension on composition in multicomponent systems may also be complicated.

5.4.1. The indium–tin–lead–bismuth system

The lateral binary systems, with the exception of the indium–tin system, re characterised by smooth surface tension isotherms [12]. As indicated in chapter 4, the concentration dependence of σ in the In–Sn system is described by the curve with a flat minimum in the range of the medium compositions. On the isothermal surfaces of σ of the ternary In–Sn–Pb [243] and In–Sn–Bi [246] systems this minimum is shown in the form of a 'depression' which is smoothed out in the presence of the atoms of the third component (lead or bismuth). The smooth surfaces transfer the dependence of σ on composition in the Sn–Pb–Bi [258] and In–Pl–Bi [246] ternary systems. Calculations show that the dependence of σ on composition in the binary and ternary systems, forming a quaternary system, is described with sufficient accuracy by a third degree polynomial. Therefore, this model was selected for the investigated quaternary system.

To calculate the coefficient of the polynomial (see [223]) we used a plan ensuring the uniform scatter of the experimental points in the three-dimensional simplex (regular tetrahedron). The points of the plan, being the nodes of the simplex lattice and used in the determination of the polynomial coefficient, are shown in Fig. 5.17.

For analysis and verification of the adequacy of the regression equation it is necessary to have data on the surface tension of four-component alloys. For this purpose, we examined the temperature dependences of σ and ρ of nine four-component melts, whose compositions are distributed on the beam section originating from the tip of the tetrahedron, corresponding to bismuth. The results and experimental data are presented in tables of the Appendix.

Table 21 (Appendix) gives the values of the coefficient of the regression equations for 623 and 773 K, and also the experimental data for the surface tension of pure metals and alloys used for the

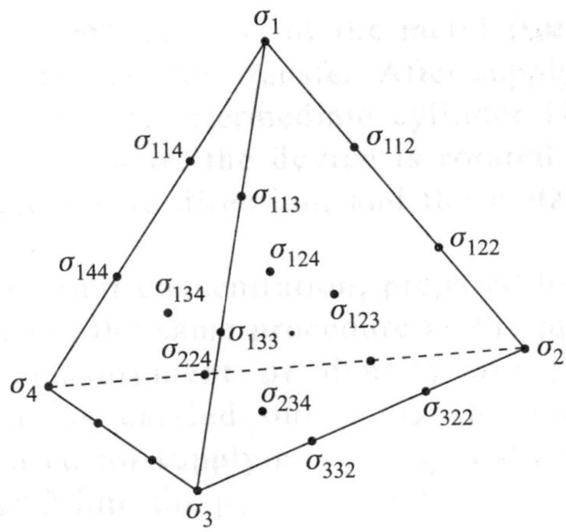

Figure 5.17. Distribution of the points of the experimental plan in the nodes of a simplex matrix for a third order model.

determination of these coefficients.

The concentration dependences of the surface tension in the discussed quaternary system have special features determined by a minimum on the isotherms σ of the indium–tin lateral binary system. The presence of the special feature increases the requirements of the verification of the adequacy of the resultant equation. Therefore, in addition to the data on the lateral binary and ternary melts, the use the experimental isotherms σ in the beam section with a constant ratio of the molar fractions of indium, tin and lead (In:Sn:Pb = 1:1:1). Comparison of the experimental data and of the isotherms calculated using the polynomials (Fig. 5.18) shows that the calculated values are in agreement with the experimental data in the measurement error range. The selected model also describes efficiently the special features of the effect of the fourth component on the 'depression' of the isothermal surface in the In–Sn–Pb and In–Sn–Bi ternary systems.

Consequently, the third order model reflects adequately the actual dependence of surface tension on the composition in the investigated quaternary system.

It should also be mentioned that the experimental values of the surface tension are in good agreement with the values calculated from different theoretical equations assuming the ideal nature of the solution. In the four-component system, this agreement is considerably better than in the ternary and binary systems forming the four-component system.

A common feature of the isothermal surfaces of σ of the lateral

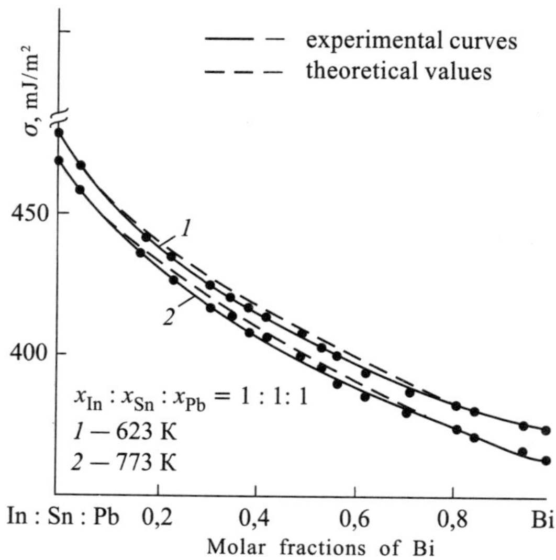

Figure 5.18. Surface tension isotherms in the indium–tin–lead–bismuth quaternary system. Solid line – the experimental curve, dashed line – theoretical values, $x_{In}: x_{Sn}: x_{Pb} = 1:1:1$; 1) 623 K; 2) 773 K.

ternary systems In–Sn–Pb and In–Sn–Bi is the presence of a 'depression' in the range adjacent to the side of the In–Sn concentration triangle. It is interesting to examine the effect of the fourth component (bismuth or lead) on this special feature of the concentration dependence of surface tension.

To illustrate the dependence of the physical–chemical properties on the composition of the multicomponent melts, we have used the isolines of the property on different planes. For example, Fig. 5.19 shows the isolines of σ of the sections, corresponding to the planes with a constant bismuth content. These graphs provide information on the effect of the bismuth content on the form of the isothermal surface of σ of the indium–tin–lead ternary system. Comparison of the graphs shows that at a low bismuth content (less than 0.05 molar fractions), the 'depression' on the isothermal surface of σ is smoothed out. A further increase of the bismuth content does not lead to any qualitative change in the form of the isothermal surface.

Lead has the same effect on the isothermal surface of σ of the indium–tin–bismuth ternary system. However, the limiting surface activity of lead in the In–Sn–Bi alloys is lower than that of bismuth in the In–Sn–Pb melts. This causes that the 'depression' on the isothermal surface of σ of the In–Sn–Bi ternary system is also retained in the presence of a fourth component (lead). The

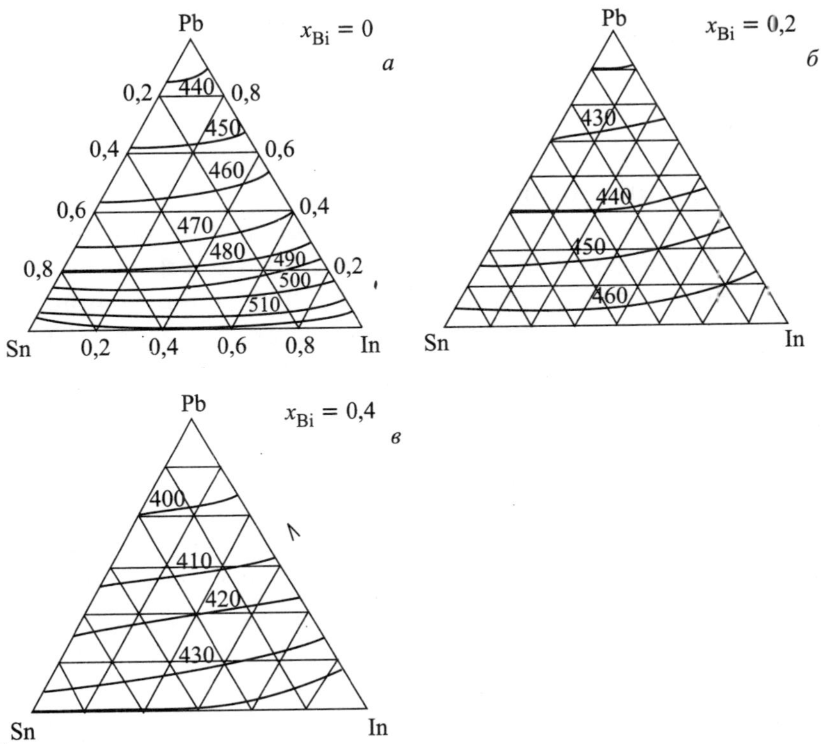

Figure 5.19. Surface tension isolines (mJ/m²) in the indium–tin–lead–bismuth system at 623 K and x_{Bi}: a) 0.0; b) 0.2; c) 0.4 molar fractions.

isothermal surfaces of σ become smooth only if the lead content exceeds 0.1 molar fractions.

Thus, with the increase of the bismuth and lead content in the indium–tin–lead–bismuth system there is a tendency for the equalisation of the relief of the isothermal surface of σ. However, with the addition of lead to the In–Sn–Bi system, the rate of the process becomes lower in comparison with the increase of the bismuth content. Taking this into account, and may be concluded that the isothermal surface σ in the factorial polygon (isosceles tetrahedron, Fig. 5.17) is anisotropic.

On the other hand, the results and theoretical calculations show that the surface tension of four-component melts is described quite accurately by the equation derived for ideal solutions. The theoretical values come closer to the experimental values with the increase of the number of components in the melt. This shows that in the systems, the surface layer of the melt approaches the ideal solution as the number of components increases.

5.4.2. The indium–lead–tin–bismuth–thallium system

The surface tension isotherms of the lead–gallium binary system are characterised by a flat minimum in the range of the medium compositions [239]. According to the literature data, the isotherms of σ of the gallium–bismuth binary system are also complicated. However, as shown in chapter 4, the concentration dependence of σ in these binary systems is described by a smooth curve, facing the composition axis by the convex side. The smooth isotherms of surface tension are also typical of other binary systems (In–Pb, In–Bi, Pb–Bi, In–Ga), forming the investigated four–component system [12].

The minimum on the isotherms of σ of the Ga–Pb binary system has a significant effect on the relief of the isothermal surfaces of σ of the Ga–Pb–In and Ga–Pb–Bi ternary systems. Calculations of the polynomial coefficients and analysis of the regression equations show that the model of the fourth order is optimum for describing the concentration dependence of σ in these systems. Therefore, this model was selected for examining the surface tension of the quaternary system.

To construct the model of the fourth order and calculate the polynomial coefficients (see [223]) it is necessary to obtain experimental data for 35 points (alloys of different compositions). The experimental data on the surface tension of these alloys at different temperatures and also the values of the polynomial coefficients are presented in Table 23 (table was taken from Goitemirov's dissertation [239]).

To analyse the resultant equation for adequacy, we examined the surface tension of five alloys of the quaternary system. Taking into account the complicated form of the concentration dependence of σ in the investigated system, reference alloys were selected to ensure that their composition is scattered uniformly throughout the entire concentration tetrahedron. This scatter enables verification of the regression equation in the entire composition range thus increasing the reliability and efficiency of the results. Comparison of the experimental values of σ at these points shows (Table 24) that the selected model of the fourth order describes with sufficient accuracy (the difference does not exceed 1%) the dependence of surface tension on composition in the investigated system.

Although the analytical representation of the dependence of surface tension on the composition of the multicomponent alloy provides less information than the graphical method of definition of the function, the analytical representation has a number of obvious advantages. The main advantage is the possibility of determination of the accurate values of the partial derivatives and, consequently, the main parameter the surface layer (adsorption, average composition, thickness of the surface layer, *etc*).

On the other hand, the resultant polynomial may be used to determine the special features of the isothermal surface of σ and also the effect of different components on the surface tension of the melt. This will be investigated on the example of gallium.

The analysis of the equation shows that gallium is an inactive element in almost the entire volume of the composition tetrahedron, with the exception of lead and alloys greatly enriched with lead (the vicinity of the tetrahedron apex, corresponding to lead). Its surface activity $G = -\left(\frac{\partial \sigma}{\partial x_{\mathrm{Ga}}}\right)_{\mathrm{In:Pb:Bi}}$ depends on the composition of the alloy and changes rapidly with the replacement of indium in the alloy by lead or bismuth. These relationships may be explained on the basis of the isotherms of σ of the lateral binary systems.

Figure 5.20 shows the isolines of σ in the quaternary system. The graph shows that the increase of the gallium content results in bending of the isolines of σ in the direction of the In–Pb concentration angle (to the edge of the In–Pb–Ga tetrahedron). On the other hand, the isothermal surface of σ shows the formation of a plateau on which the replacement of lead by bismuth has no effect on σ:

$$\left(\frac{\partial \sigma}{\partial x_{\mathrm{Bi}}}\right)_{X_{\mathrm{Ga}}, X_{\mathrm{In}}} = \left(\frac{\partial \sigma}{\partial x_{\mathrm{Pb}}}\right)_{X_{\mathrm{Ga}}, X_{\mathrm{In}}} = 0$$

Thus, the phenomenon of concentration buffering is typical not only ternary but also multicomponent systems. As mentioned previously, in the ternary systems this phenomenon has been detected many times.

As far as we know, this phenomenon has been detected for the first time in the four-component system. These questions will be examining in greater detail in the following section.

The surface tension isotherms of the lead–gallium binary system are characterised by a flat minimum in the range of the medium compositions. As in the In–Sn–Pb–Bi quaternary system, on the isothermal surfaces of σ this minimum is reflected in the form of a 'depression'. The depth of the 'depression' decreases with the increase of the content of the third or fourth component, and at a

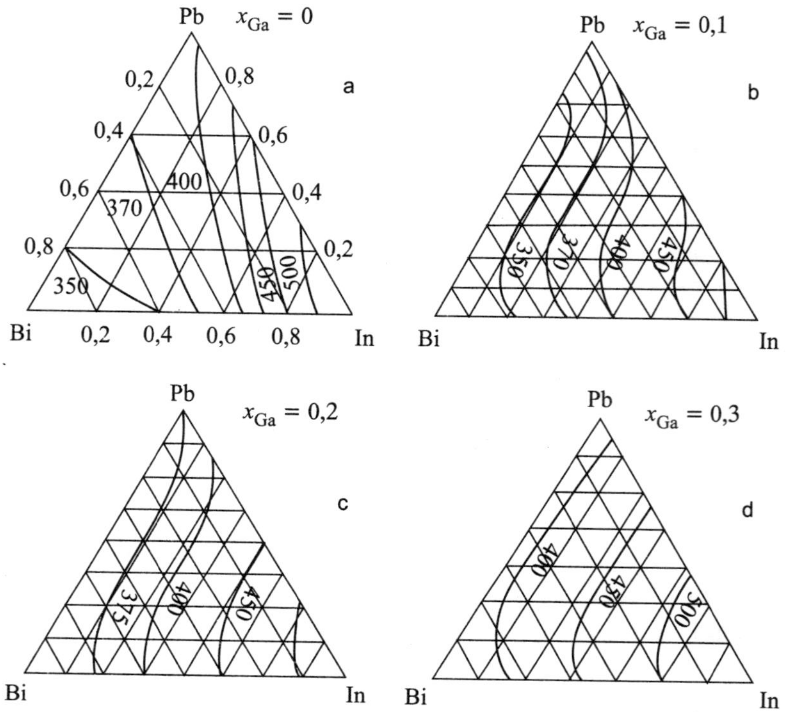

Figure 5.20. Surface tension isolines (mJ/m²) at 1073 K in the indium–bismuth–lead–gallium system. x_{Ga} =: a) 0.0; b) 0.1; c) 0.2; d) 0.3 molar fractions.

specific total concentration of these components the isothermal surface of σ becomes smooth. This region is also characterised by the same relationships as the previous region. Calculations show that a decrease of the minimum of the isotherms of σ of the Ga–Pb binary system as a result of the addition of bismuth is considerably greater in comparison with the addition of indium. This is reflected in both ternary and quaternary systems and is caused by the fact that bismuth, in contrast to indium, is a surface active component. Similar special features of the effect of components on the form of the isotherms of σ have been detected previously in the examination of ternary systems. Consequently, it may be assumed that these relationships are also typical of the multicomponent systems.

Thus, the dependence of surface tension on composition in the investigated four-component system is complicated. However, as in the ternary systems, the special features of the concentration dependence of σ are linked with the isothermal surface tension of lateral binary systems. The isotherms of σ in the beam sections, originating from the apex of the tetrahedron, do not contain any qualitatively new special features in comparison with the isotherms of σ of the lateral binary systems.

5.5. Special features of the variation of the physical and chemical properties with the composition of the multicomponent melt. Concentration buffering of surface tension

The previously presented experimental data on the surface tension and density of seven ternary and two quaternary systems may be used to define some relationships governing the variation of surface tension and molar volumes with the composition of the multicomponent melt. In particular, the relationships of the effect of the third and fourth component and also the combined effect of these components on the special features of the surface tension isotherms of the binary melts have been revealed. This was possible owing to the fact that in all the examined multicomponent systems, with the exception of In–Bi–Tl, the isotherms of σ of one of the lateral systems (In–Sn or Tl–Pb) are characterised by a flat minimum in the range of the medium compositions. The third component is represented by both the surface active metals lead (In–Sn–Pb); bismuth (Tl–Pb–Bi); thallium (In–Sn–Tl), and by the inactive metals, *i.e.*, gallium (In–Sn–Ga), indium (Tl, Pb–In), thallium (Tl–Pb–Sn).

A general feature of the isothermal surface of σ of the investigated alloys is the presence of a depression in the region of the compositions adjacent to these binary systems. On the isotherms of σ of the sections with a constant content of the third component, and also on the isotherms of binary melts, the 'depression' is manifested in the form of a minimum.

It has been established that the increase of the content of the third component in the section results in gradual smoothing of the special features (minimum) on the surface tension isotherms. The intensity of the effect of the third component is directly linked with the value of its limiting surface activity.

Thus, according to the results, the minimum on the isotherms of the In–Sn and Tl–Pb binary systems disappears at the following concentrations of the third component: 0.1 lead (In–Sn–Pb); 0.2 bismuth (Tl–Pb–Bi); 0.7 thallium (In–Sn–Tl); 0.4 gallium (In–Sn–Ga); 0.4 lead (Tl–Pb–Sn); 0.6 indium (Tl–Pb–In). In this series, the systems are distributed in the order of decreasing limiting surface activity of the third component in the alloys of the binary system. If in the first three systems the third component is surface active, then in the remaining systems it is inactive. These results show clearly the validity of the established relationship. In fact, as the limiting surface activity of the component increases, the minimum of the isotherms of σ disappears at a lower concentration of the active component, *i.e.* the effect of the third component on the special features of the concentration dependence of surface tension becomes stronger. An exception is the In–Sn–Tl system in which the 'depression' is noticeable up to 0.7 molar fractions of thallium regardless of the capillary activity of thallium in the indium–tin alloys.

The detected effect of the third component may be explained by the fact that its content in the surface layer is determined (see 1.5) by the derivative $\left(\frac{\partial\sigma}{\partial x_3}\right)_{K_{12}}$ which also characterises surface activity. Calculations of the composition of the surface layer of these melts, carried out in the following section, show that the 'depression' on the isothermal surface of σ is smoothed-out at approximately the same concentrations of the third component the surface layer (at $x_3 \approx$ 0.4–0.5 molar fractions). In the four-component systems, as in the ternary systems, the effect of the third and fourth component on the isotherms of σ is determined by the values of their limiting surface activities:

$$G_{0i} = -\lim_{x_i \to 0}\left(\frac{\partial\sigma}{\partial x_i}\right)_{\frac{x_i}{x_r}}$$

No interaction between the admixtures has been detected. The combined effect of the third and fourth component on the special features of the isotherms of σ of the binary system is determined by their total concentration in the melt, *i.e.*, it is added up in accordance with the additive law. Evidently, this relationship is caused by the relatively low capillary activity of the component in the investigated systems and by the absence of strong chemical bonds between them.

The molar volumes of the investigated melt characterised by a positive deviation from the additive values, and in the In–Sn–Ga and In–Pb–Bi–Ga melts coincide with these values in the experimental error range. The isotherms of the molar volumes in the beam sections of the ternary and quaternary systems have positive deviations from the additive straight lines. However, in none of the investigated systems is the absolute value of these deviations higher than the deviations of the molar volumes from the additivity in the lateral binary systems.

The concentration dependence of the surface tension of the investigated multicomponent systems is complicated. However, the special features of the isothermal surfaces of σ are directly linked with the isotherms in the lateral binary systems. The isotherms of σ and V- beam sections of the multicomponent system do not contain any qualitatively new special features in comparison with the isotherms of these properties in the lateral binary systems.

The most interesting special feature of the concentration dependence of σ of the multicomponent solutions is the so-called concentration buffering. This phenomenon was detected in our work on the In–Bi–Tl and In–Pb–Tl ternary systems and also the In–Pb–Bi–Ga quaternary system. In this phenomenon, the multicomponent system contains sections in which the variation of composition has no effect on surface tension. Since the variation of the composition of the In–Pb–Tl melts in the section $x_{In}: x_{Tl} = 1:1$, and In–Bi–Tl $x_{In}: x_{Bi} = 1:2$ has, according to the results, no effect on the surface tension of the melt, the value of σ is independent of the content of lead or thallium.

The phenomenon of concentration buffering of σ has also been detected in the In–Pb–Bi–Ga system. Actually, the experimental data show that in some region of the compositions, the replacement of lead by bismuth has no effect on surface tension:

$$\left(\frac{\partial \sigma}{\partial x_{Bi}}\right)_{X_{Ga}, X_{In}} = \left(\frac{\partial \sigma}{\partial x_{Pb}}\right)_{X_{Ga}, X_{In}} = 0$$

It may easily be seen that this condition differs qualitatively from those for the discussed ternary systems.

As mentioned previously, the phenomenon of concentration buffering was detected a long time ago and examined theoretically in [6, 13, 19, 251]. However, the analysis of these investigations shows that the literature does not contain any unambiguous information on the definition of the phenomenon of concentration buffering of surface tension. This is associated with the existence of a large number of possible mechanisms of the variation of the composition of the multicomponent solution. This also determines the difference in the phenomena which we investigated in the ternary and quaternary systems.

Semenchenko and Popel' examined a case of the variation of composition along lines with a constant content of a surface-active component. The buffering section is the one satisfying the condition:

$$\left(\frac{\partial\sigma}{\partial x_i}\right)_{x_j}=0 \tag{5.5.1}$$

where x_j is the molar fraction of the surface active component.

This case was detected in our work in the In–Pb–Bi–Ga quaternary system (see paragraph 5.4). Rusanov investigated the second method of the variation of the composition of the ternary system and he named the buffering section as the section of the concentration angle for which the following equation is satisfied:

$$\left(\frac{\partial\sigma}{\partial x_i}\right)_{K_{ir}}=0 \tag{5.5.2}$$

where $K_{ir} = x_i/x_r$ is the ratio of the molar fractions of the components.

Equations (5.5.1) and (5.5.2) are not only independent but in the real solutions cannot be fulfilled at the same time. In fact, these conditions are generally identical with the requirement that the value of σ should be independent of the composition of the ternary solution. On the other hand, the fulfilment of these conditions for the given composition indicates mathematically the presence of a 'triple' extreme (minimum or maximum).

Thus, the literature does not contain any unambiguous definition of the phenomenon of concentration buffering and this creates certain difficulties in understanding the nature of this phenomenon and results in misunderstanding and objections. In most cases, the phenomenon of concentration buffering has been associated with the special features of the structure of the alloys. However, it is quite easy to show that this phenomenon is also typical of ideal systems [47]. At least, it holds for the case examined in [19].

In fact, let us assume that the concentration dependence of σ is described by the equation for ideal solutions:

$$\sigma=\sigma_1-\frac{RT}{\omega}\ln\left[x_1+x_2\exp\left(\frac{\sigma_1-\sigma_2}{RT}\omega\right)+x_3\exp\left(\frac{\sigma_1-\sigma_3}{RT}\omega\right)\right] \tag{5.5.3}$$

The buffering secant is characterised by the condition (5.5.2). Substituting the expression for σ from (5.5.3) into (5.5.2) gives:

$$1+K_{23}-K_{23}\exp\left(\frac{\sigma_1-\sigma_2}{RT}\omega\right)-\exp\left(\frac{\sigma_1-\sigma_3}{RT}\omega\right)=0 \qquad (5.5.4)$$

This equation leads to:

$$K_{23}=\frac{\exp\left(\frac{\sigma_1-\sigma_3}{RT}\omega\right)-1}{\exp\left(\frac{\sigma_2-\sigma_1}{RT}\omega\right)-1} \qquad (5.5.5)$$

so that it is possible to determine the position of the buffering secant which is determined by the value of the ratio $K_{23} = x_2/x_3$.

If it is taken into account that according to the physical meaning $K_{23} > 0$, then (5.5.5) gives the following conditions essential for the appearance of concentration buffering $\sigma_2<\sigma_1<\sigma_3$ or $\sigma_2>\sigma_1>\sigma_3$.

Consequently, the concentration buffering appears in the ternary system if the system is formed by components where one of the components is surface-active and the other component is surface-inactive in a binary system with the first component. Thus, the phenomenon of concentration buffering may also be found in ideal solutions.

When analysing the phenomenon of concentration buffering, we investigated the method of changing the composition in the beam sections of the concentration triangle.

Consequently, the conclusions made previously, also apply to this case. One of the main tasks of the theory of surface phenomena in melts is the theoretical description of the dependence of surface tension on composition and temperature, which is equivalent to the derivation of the equation of state of the surface layer. Regardless of the large number of proposed equations, this problem has not been completely solved. The task of collecting and theoretical justification of the experimental data for the surface tension of binary and, in particular, multicomponent melts is also important.

We have calculated the value of σ of the experimentally investigated melts using the equations derived by Zhukovitskii [80] and Popel'–Pavlov [92]. Because no data are available on the coefficients of thermodynamic activity, calculations were carried out assuming that the solutions are ideal. Comparison of the theoretical isotherms with the experimental values shows that the Popel'–Pavlov equation describes the experimental curves with the accuracy of 3–5%. However, this equation does not describe the minimum on the isotherms of σ of the indium–tin and thallium–lead binary systems and the isothermal surface of σ in the regions adjacent to these binary systems. The isotherms, obtained using the equation proposed by Zhukovitskii (5.4), are 10–50% higher than the experimental values. Thus, the Popel'–Pavlov simulation-thermodynamic equation describes quite efficiently the concentration dependence of the surface tension in the discussed multicomponent alloys, irrespective of the assumption on the ideal nature of the solutions. It should be mentioned that the differences between the calculated values of σ and the experimental values in multi-component melts are smaller in comparison with lateral binary melts. For accurate calculations using this equation, it is necessary to obtain data on the molar and partial–molar surfaces, and also the coefficients of thermodynamic activity of the components in multicomponent melts. These circumstance complicate and delay of the application of the equation in the analysis of surface phenomena in multicomponent melts. Therefore, we have proposed semi-empirical formulae for calculating the surface tension of multi-component melts on the basis of the properties of the lateral binary systems. The equations and calculations carried out using these equations will be discussed in greater detail in chapter 6.

6

Thermodynamic parameters of the surface layer of multicomponent melts and methods of predicting the parameters

In the first two chapters, attention was given to the main thermodynamic equations of the interface with special reference to multicomponent two-phase systems. Equations were derived and a method developed for calculating both the Gibbs excess quantities and in various properties of the surface layer, as the composition averaged with respect to the thickness, effective thickness, molar and partial–molar surface of the component. Calculations using the equations require data on the temperature and concentration dependence of surface tension, molar volumes and chemical potentials of the component in the melt. In chapter 5, we presented the experimental data on the surface tension and density of seven ternary and two four-component systems.

Using these data, the equations mentioned previously are used in the present chapter to calculate the dynamic parameters of the surface layer of these melts. In the first paragraph, in terms of the Gibbs excess quantities, we discuss the adsorption processes in these melts, and in the second section the processes are analysed on the basis of the method of the finite thickness layer.

In addition to this, the results of our experimental investigations and analysis of the literature data are used to determine the relationship of the concentration dependence of the surface properties of multicomponent solutions with the isotherms of these properties for the appropriate binary systems. Various methods of predicting the surface tension of the multicomponent melt using the data on the binary systems are described.

6.1. Adsorption phenomena in multicomponent metallic melts

In this section, we discuss the results of calculating the adsorption of components in the investigated ternary and quaternary systems. Of the variants, investigated in chapter 1, we selected the equimolar dividing surface, i.e. *N*-variant adsorption according to Guggenheim and Adam. In the two-phase multicomponent liquid–vapour systems away from the critical temperature, the adsorption on this surface is linked by a simple equation with the average composition of the surface layer.

To calculate the thermodynamic parameters of the surface layer, in addition to the data on the surface tension and density, it is necessary to have data on the coefficients of thermodynamic activity of components in a multicomponent melt. For the majority of multicomponent melts no such data are available in the literature. Therefore, the values of the thermodynamic activity coefficients were estimated on the basis of data on the lateral binary systems, calculations were carried out assuming that the multicomponent solutions are ideal, *i.e.*, the thermodynamic activity coefficients of the component were assumed to be equal to unity.

It is evident that this approximation is not always justified and requires preliminary examination.

To evaluate the degree of deviation of the solutions from the ideal state and its effect on the form of the isotherms of the surface properties of the multicomponent systems, the calculations of the adsorption of components in indium–tin–lead and thallium–lead–bismuth systems were carried out using the equations for both the real and ideal solutions. The values of the coefficient of thermodynamic activity were calculated using the Koller method [252] which gives satisfactory results in multicomponent systems in which one of the binary systems is close to ideal (In–Sn or Tl–Pb), and others are characterised by positive deviations from the Raoult law (In–Pb, Sn–Pb and Pb–Bi). The applicability of the method for calculating activities in the discussed system has also been shown in [205, 206].

The results of the calculations show that in these systems the effect of the non-ideal state of the solutions on the position of the adsorption isotherms does not exceed the limits of the error of determination of adsorption. It should be mentioned that the error of determination of adsorption reaches 10%. Consequently, at not too large deviations of the lateral binary systems from the Raoult's law, the model of the ideal solution may be used for calculating and analysing the adsorption phenomena in multicomponent melts. Therefore, the adsorption in these systems was calculated assuming that the solutions are ideal.

As mentioned in chapter 5, the concentration dependence of the surface tension of certain ternary systems is close informed. Consequently, the similar dependences on the composition are also characteristic of the adsorption of components in these melts. As regards the similarity of the isothermal surfaces of adsorption, of the investigated ternary systems may be divided into two groups.

The first group includes systems in which the said component is represented by a surface-active metal: indium–tin–lead, thallium–lead–bismuth and indium–tin–thallium.

The second group includes ternary systems indium–tin–gallium, thallium–lead–indium and thallium–lead–tin. In contrast to the first group, in these melts, the third component is inactive both on pure metals and on all melts of the lateral binary system.

The bismuth–thallium–indium ternary system differs from the previously examined systems by the smooth isothermal surface σ and by the absence of special features on the isotherms of surface tension of lateral binary systems. Consequently, the adsorption surfaces in this system are relatively simple.

6.1.1. Effect of the surface-active component on the adsorption processes in binary melts

The concentration dependence of the adsorption of components in multicomponent systems may be graphically represented in the form of isolines (Figs. 6.1 and 6.2). However, because of the complicated relief of the isothermal surfaces of adsorption, the perception and analysis of the isolines is associated with certain difficulties. A more complete and convincing information on the adsorption processes in multicomponent melts is provided by the adsorption isotherms in different sections. Therefore, the results of calculation of adsorption in the investigated systems are presented mainly in the form of isotherms in the beam sections.

In the In–Sn–Pb, In–Sn–Tl and Tl–Pb–Bi ternary alloys the

isotherms of σ of the indium–tin and thallium–lead binary systems characterised by the formation of a flat minimum, and the added metals (lead, thallium and bismuth) are surface active. This results in the same nature of the isothermal surfaces σ and the adsorption of components in these systems. The adsorption isotherms of the surface-active components in the beam sections are shown in Figs. 6.3 and 6.4.

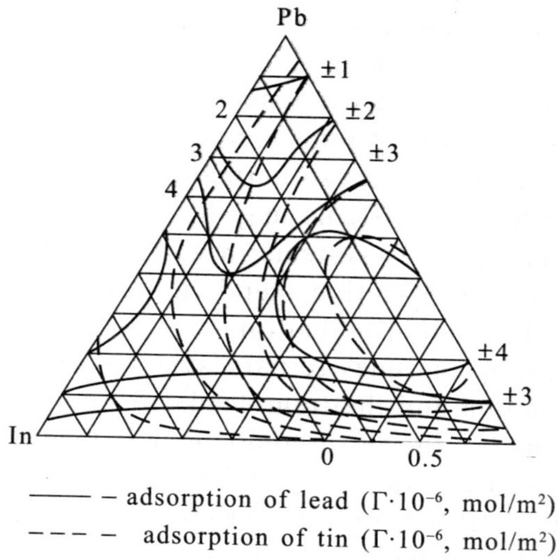

Figure. 6.1. Adsorption isolines in the indium–tin–lead ternary system at 773 K.

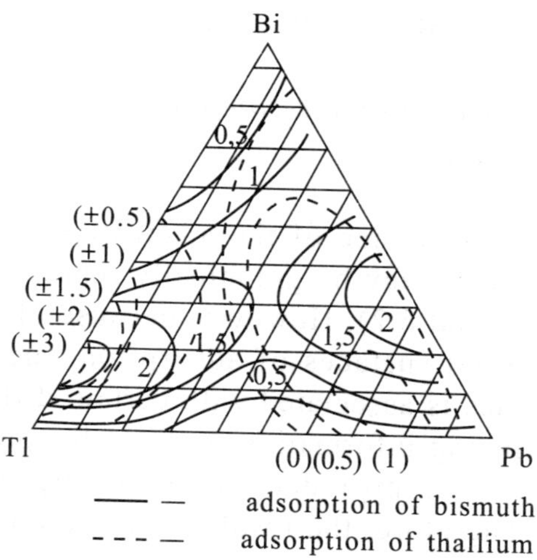

Figure. 6.2. Adsorption isoline (10^{-6} mol/m^2) in the thallium–lead–bismuth system at 773 K.

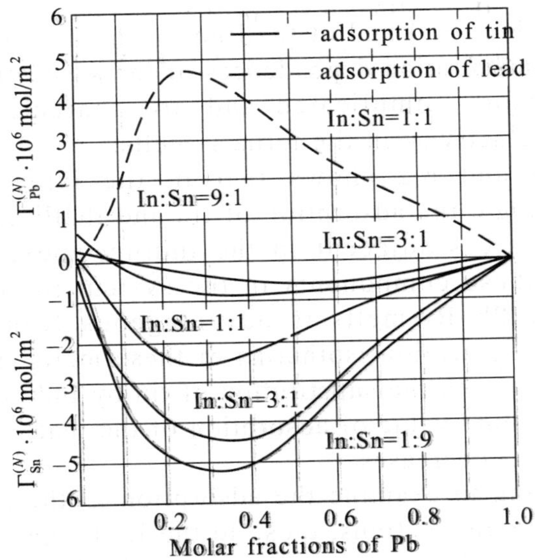

Figure. 6.3. Adsorption of components in indium–tin–lead melts at 773 K.

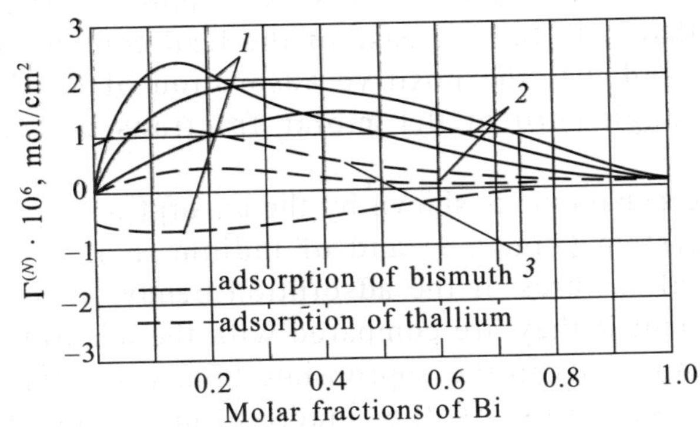

Figure. 6.4. Adsorption of components in thallium–lead–bismuth melts at 773 K.

As indicated by Figs. 6.3 and 6.4, the concentration dependence of the adsorption of the surface-active metals in the beam sections of the concentration angle have similar form in comparison with the adsorption isotherms in the lateral binary systems.

For example, lead in the indium–tin–lead ternary system is characterised by positive adsorption both on pure metals like indium and tin and on all In–Sn alloys. The isotherms of adsorption of lead in the sections with the constant ratio of indium to lead have the form similar to the isotherms in the indium–lead and tin–lead binary systems.

The behaviour of the adsorption isotherms of bismuth in the Tl–

Pb–Bi system (Fig. 6.4) and of thallium in the In–Sn–Tl system is similar. The concentration dependences of the adsorption of tin and thallium are more complicated and are characterised by the variable surface activity in the ternary melts.

Figures 6.1 and 6.2 shows that in the large part of the concentration angle the adsorption of tin and thallium is negative. Only in small areas adjacent to the indium apex, the positive adsorption of these components in the system In–Sn–Pb and of lead in the Tl–Pb–Bi melts is advantageous from the energy viewpoint. The adsorption isotherms of these melts (Figs. 6.3 and 6.4) indicate that with the substitution of tin by indium or thallium by lead, the absolute value of adsorption of the inactive component (tin and thallium) decreases.

It is interesting to examine the adsorption isotherms of tin and thallium in the beam sections, close to the In–Pb and Pb–Bi sides. These isotherms indicate the suppression of adsorption of tin or thallium with the increase of the content in the melt of the more surface active components, *i.e.* lead or bismuth. For example, Fig. 6.3 shows that with the increase of the lead content in the section x_{In}: x_{Sn}=3:1 and 9:1, the positive adsorption of tin decreases and passing through zero at 0.1 molar fractions of lead, becomes negative.

The same behaviour is shown by the adsorption of thallium in the Tl–Pb–Bi system (Fig. 6.4) and of indium in In–Sn–Tl systems. These special features of the adsorption behaviour of components become evident if they are compared with the adsorption isotherms of these components in the appropriate binary systems. In fact, the adsorption isotherms of the inactive metal in the sections, corresponding to the constant ratio of the molar fractions of other components, occupy an intermediate position between the isotherms of its adsorption in the appropriate binary systems.

6.1.2. Adsorption behaviour of the inactive component in In–Sn–Ga, Tl–Pb–In, Tl–Pb–Sn and Tl–Bi–In ternary melts

As mentioned previously, the second group is represented by the indium–tin–gallium, thallium–lead–indium and thallium–lead–tin ternary systems. The common feature of these melts, like of the melts of the first group, is the presence of a minimum on the isotherms of σ of one of the lateral systems, In–Sn or Tl–Pb. However, in contrast to the melts of the first group, in these systems the third component (gallium, indium, tin) increases the surface tension of the appropriate pure metals and their alloys.

Although indium is inactive in the Tl–Bi alloys, the Tl–Bi–In ternary system differs from others by the absence of special features on the isotherms of σ of the lateral binary systems. Therefore, the adsorption phenomena in these systems are investigated in a separate section.

The results of calculation of the adsorption of components in the discussed ternary systems are shown in Figs. 6.5–6.10.

The graphs indicate that the adsorption of active components (gallium in the In–Sn–Ga system, tin and indium in the Tl–Pb–In, Tl–Pb–Sn and Tl–Bi–In system) is negative over the entire area of the concentration angle. The adsorption isotherms of these components in the beam sections are similar to the isotherms of the appropriate binary systems. The concentration dependence of the adsorption of surface active components with the variation of the composition of the components in the same sections is slightly more complicated. As indicated by Figs. 6.5 and 6.7, the adsorption of tin in the In–Sn–Ga system and of thallium in the Tl–Pb–Sn melts is negative in the alloys enriched with these metals.

With an increase of the content of the inactive component (gallium in the In–Sn–Ga system, tin in the Tl–Pb–Sn melts), the adsorption isotherms of tin and thallium pass through zero and become positive. Thus, these systems are characterised by a process inverse to the phenomenon of suppression of adsorption.

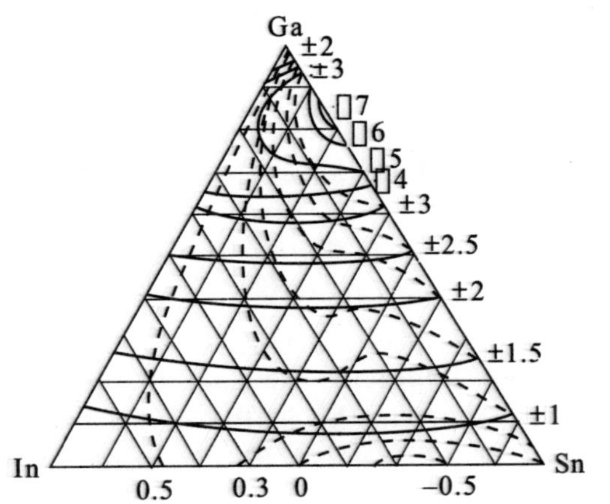

——— — adsorption of gallium (10^{-6} mol/m^2)
- - - - — adsorption of tin

Figure. 6.5. Adsorption isolines (i) (in 10^{-10} mol/cm^2) in the In–Sn–Ga system at 250°C.

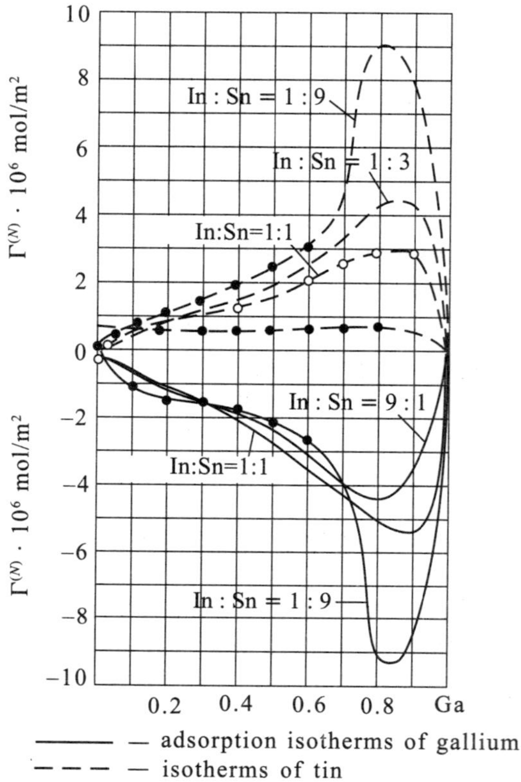

Figure. 6.6. Isotherms of adsorption of components in the indium–tin–gallium system at 523 K.

The increase of the content of the inactive component in the melts (gallium or tin) resulted in positive adsorption and, consequently, in the displacement into the surface layer of the atoms of less inactive components tin or gallium in the In–Sn–Ga and Tl–Pb–Sn ternary systems, respectively.

Figures 6.6 and 6.7 also show that with the decrease of the fraction of the component in the section, the absolute value of its adsorption decreases and ceases to depend on the content of the inactive component. This is clearly indicated by the adsorption isotherms of tin in the section $x_{In}:x_{Sn}$ = 9:1 (Fig. 6.6). The graph shows that the adsorption of tin in the wide range of the compositions is independent of the gallium content. An identical phenomenon was observed on the isotherms of thallium in the beam sections $x_{Tl}:x_{Pb}$ = 1:9 of the Tl–Pb–Sn and Tl–Pb–Sn ternary systems (Fig. 6.7).

As mentioned previously, the thallium–bismuth–indium ternary melts differ from the previously examined melts by the absence of

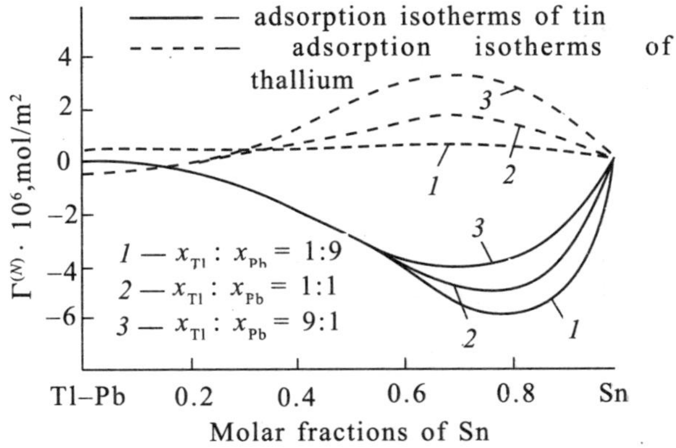

Figure. 6.7. Adsorption isotherms of components at 623 K in the thallium–lead–tin binary system.

special features on the surface tension isotherms. This also resulted in a change of the adsorption of inactive indium in these melts (Fig. 6.10). The adsorption isotherms in the beam sections are similar to the isotherms in the appropriate binary systems. This dependence is more evident in Figs. 6.8 and 6.9 where one graph combines the adsorption isotherms in binary and ternary systems.

In this case, a component which shows surface activity in one metal and is inactive in the other metal was selected. Consequently, this makes it possible to examine the dynamics of the variation of the type of adsorption isotherms. Actually, in the In–Bi–Tl ternary system, thallium shows high surface activity on indium, but is inactive on bismuth. Its behaviour in the Pb–In–Tl melts is similar.

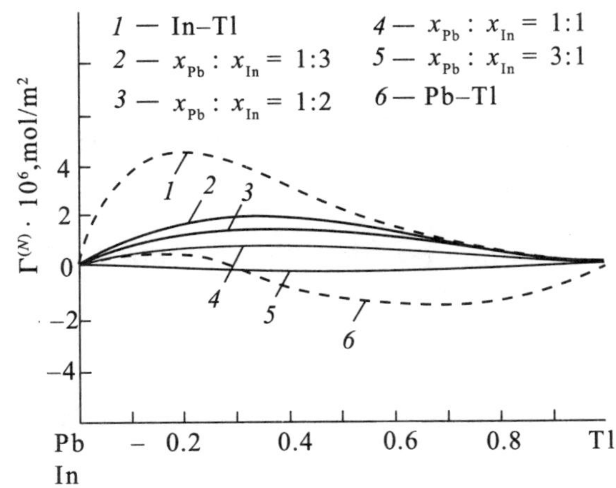

Figure. 6.8. Adsorption isotherms of thallium at 633 K of the thallium–lead–indium ternary system.

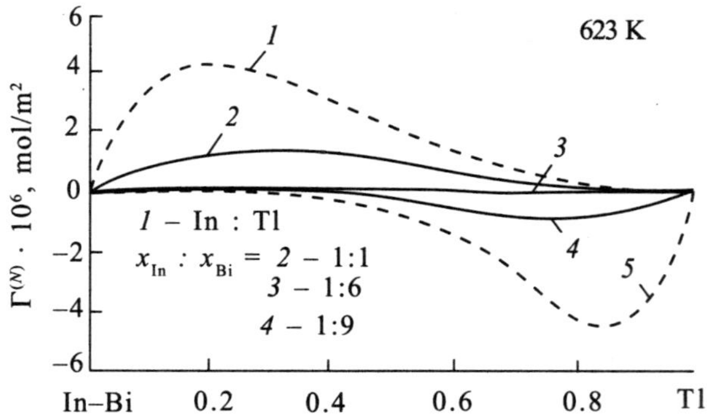

Figure. 6.9. Adsorption of thallium in the thallium–bismuth–indium system at 623 K.

The graphs indicate that the adsorption isotherms of thallium in the sections of the ternary systems occupy an intermediate position between the isotherms in the appropriate binary systems. With the replacement, in the section, of the atoms of one component by the atoms of another component, the isotherms gradually transfer from one binary system to another.

Thus, the adsorption phenomena in the investigated ternary systems are complicated. The following phenomena were detected: suppression of the adsorption of the component by the adsorption of another more active component; concentration buffering when the adsorption of one of the components is equal to zero; the phenomenon characterised by the independence of the values of the adsorption of the surface active component on the content of the inactive component.

In the ternary systems, the special features are determined directly by the adsorption behaviour of the components in the appropriate lateral binary systems, *i.e.*, the adsorption processes in the ternary melts may be predicted on the basis of the adsorption isotherms in the lateral systems.

We examine from this position the adsorption phenomena in the investigated four-component systems.

6.1.3. Adsorption phenomena in indium–tin–lead–bismuth and indium–lead–gallium–bismuth four-component solutions

The dependence of surface tension on the composition of the four-component melts is described by a polynomial. This greatly facilitates the calculation of the adsorption of components in these

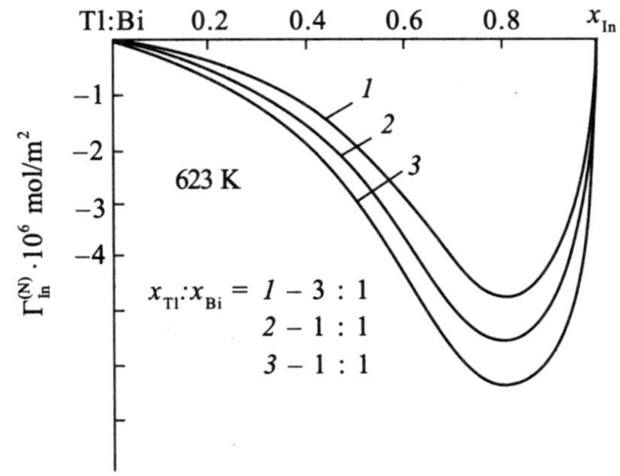

Figure 6.10. Adsorption of indium in the thallium–bismuth–indium system at 623 K.

melts because it is possible to determine the partial derivatives of the surface tension for any direction by simple differentiation. The form of the polynomial and the values of its coefficient presented in the Appendix. Adsorption was calculated assuming that the solution is ideal.

As shown in chapter 1, if the dependence of surface tension on the composition of the melt is available, there are no major difficulties in adsorption calculations. However, with the increase of the number of components analysis becomes far more complicated and it is also more difficult to describe adsorption. If its dependence on the composition of the ternary melts may be presented in the three-dimensional space, then adsorption in the four-component system is a function of three variables. The isolines of the properties, used in these cases, are also not very effective because of the complicated form of the concentration dependence of sorption. Therefore, Figs. 6.11 and 6.12 show the isotherms of adsorption of components with the variation of the composition in the beam section of the tetrahedron of the compositions.

The results of the calculations show that the concentration dependence of adsorption in In–Sn–Pb–Bi and In–Bi–Pb–Ga melts is complicated. In particular, there are phenomena such as the suppression of adsorption of some components by others and the phenomenon of concentration buffering. However, as in the ternary systems, these special features are directly linked and also determined by the adsorption processes in binary systems, forming the multicomponent system. As an example, Figs. 6.11 and 6.12 showed the isotherms of adsorption of bismuth in In–Sn–Pb–Bi and of indium in In–Bi–Pb–Ga melts. The components were selected to

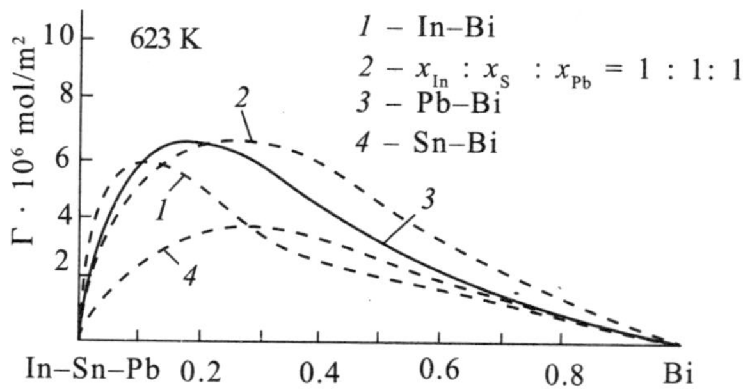

Figure 6.11. Adsorption of bismuth in indium–tin–and–bismuth melts.

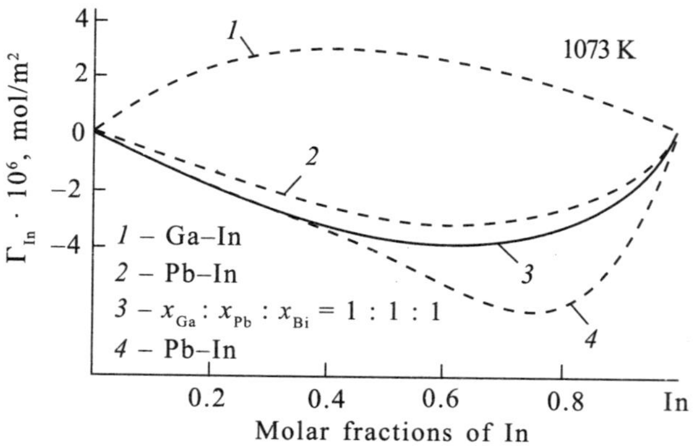

Figure 6.12. Adsorption of indium in the indium–lead–bismuth–gallium system at 1073 K.

ensure that they differ in the surface activity. Bismuth shows surface activity in both pure metals indium, tin, lead, and in the alloys of the quaternary system. In contrast to bismuth, in the, which increases the surface tension of lead, bismuth and their alloys, shows high surface activity on gallium and on gallium-enriched alloys.

For comparison, in addition to the adsorption isotherms in the sections of the quaternary systems, the figures also show the concentration dependences of adsorption in the lateral binary systems. The graphs indicate that the isotherms of bismuth and tin in the beam sections have similar form and occupy an intermediate position between their adsorption isotherms in lateral binary systems.

6.2. Composition and thickness of the surface layer of multicomponent melts

In the method of the finite layer thickness, the condition of the surface is also characterised by parameters such as composition, effective thickness, molar and partial–molar surfaces. The direct experimental determination of these quantities is associated with insurmountable difficulties. Therefore, it is important to develop methods of determination of these quantities from the available or easily measurable properties of the melts.

The property of the surface of the liquid which is easy to determine with high reliability is the surface tension. Therefore, in chapter 2, the fundamental thermodynamic equations for the flood surface layer was processed to derive equations which can be used to calculate, from the data on the concentration dependence of surface tension, the molar surface, composition and thickness of the surface layer of multicomponent melts.

In calculations using these equations it is not necessary to make any *apriori* assumptions regarding the concentration dependence of the molar surface and the thickness of the surface layer. These dependences are also determined from the concentration dependence on the surface tension of melts.

Using the procedure described in chapter 2, we calculated the thermodynamic parameters of the surface layer of ternary and quaternary systems investigated in our experiments. The results are discussed below.

The thermodynamic activities of the components, required for the calculations, in the indium–tin–lead and thallium–lead–bismuth ternary melts are determined by the Koller method. In other systems, calculations are carried out assuming that the solution is ideal. The values of the molar surfaces and of the thickness of the surface layer of the pure components, included in the calculation equation, are presented in section 4.1.

As regards the type of the isothermal surfaces of the discussed properties, the investigated ternary systems are similar. Therefore, the figures show the most interesting and typical curves indicating the general relationships of the variation of the surface properties with the composition of the solution and the surface layer.

Figures 6.13–6.16 show the isotherms of the molar fractions of the third component (lead, bismuth and gallium) in the surface layer of the In–Sn–Pb, Tl–Pb–Bi and In–Sn–Ga melts, respectively. Lead and bismuth are surface–active components, and gallium is inactive.

The isotherms of the molar fractions of lead and bismuth in the surface layer of the indium–tin–lead and thallium–lead–bismuth melts (Figs. 6.13, 6.14) show that the surface of the liquid is enriched with these components. The maximum differences in the

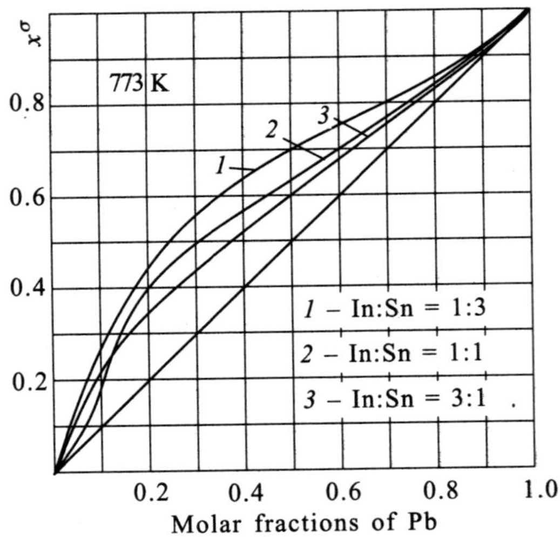

Figure 6.13. The lead content of the surface layer of indium–tin–lead melts at 773 K.

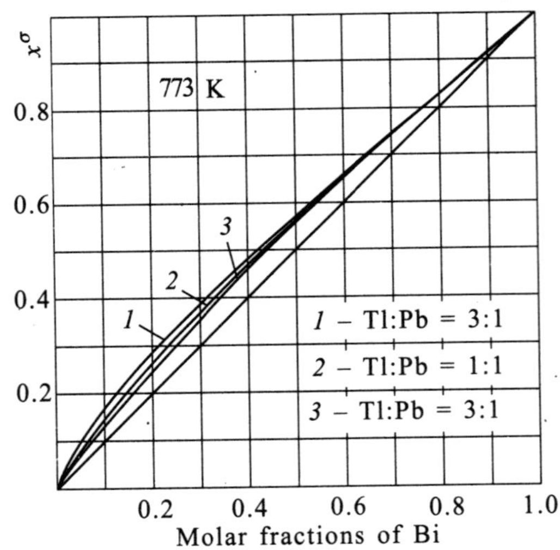

Figure 6.14. The bismuth content of the surface layer of thallium–lead–bismuth melts at 773 K.

composition of the surface layer and the volume of the melt are found in the range of low concentrations of the surface-active component (0.2~0.3 molar fractions).

The isotherms x_m^σ in the beam sections of the ternary systems are similar in form and occupy an intermediate position between the isotherms of the content of this component in the appropriate lateral binary systems. A similar behaviour is shown by the surface active components in other investigated ternary systems (thallium in In–Sn–Bi melts and bismuth in In–Pb–Bi melts).

The behaviour of the inactive component in the surface layer of the melts may be examined on the example of gallium in the indium–tin–gallium ternary system. The behaviour of indium and tin in the Tl–Pb–In and Tl–Pb–Sn system is similar to that of gallium.

The composition of the surface layer of the In–Sn–Ga melts in the form of the function of the molar fractions of gallium in the volume phase is shown in Fig. 6.15. The graph indicates that the content of gallium in the surface layer is lower than in the volume of the melt. The maximum deviation is recorded at 0.9 molar fractions of thallium.

The isotherms of the molar fraction of thallium in the surface layer in the beam sections of the ternary system and indium–gallium and tin–gallium binary melts have similar form.

In addition to the isotherms x^{σ}_{Ga}, Fig. 6.15 shows the dependences of the molar fractions of tin on the gallium content of the melts with the variation the composition in the same beam sections (x_{In}:x_{Sn}= const). An interesting special feature of these curves is that at 0.05 molar fractions of gallium they intersect with the additive straight lines, *i.e.*, tin which is inactive in the indium–tin alloys, becomes surface-active.

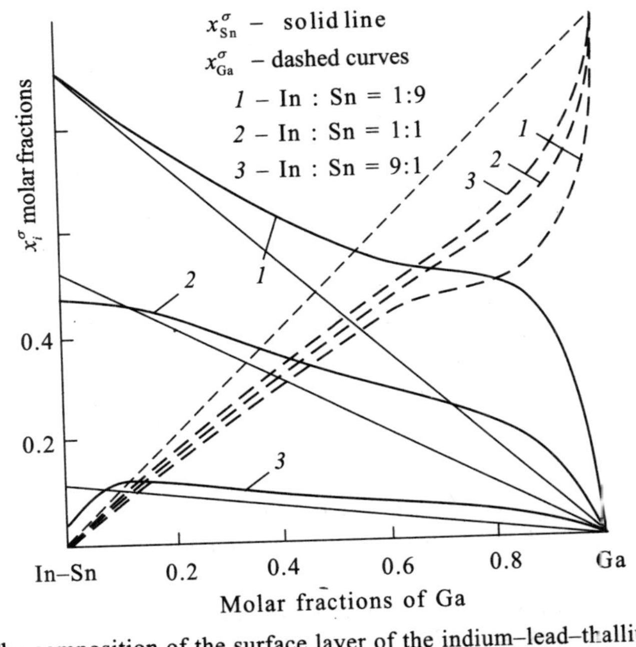

Figure 6.15. The composition of the surface layer of the indium–lead–thallium melts at 523 K.

In fact, if the tin content of the surface layer at a low gallium concentration (less than 0.05 molar fractions) is lower than in the volume, with the increase of the amount of gallium in the melt, gallium displaces tin into the surface layer. This shows that the presence of a third or fourth component may change qualitatively of the composition of the surface layer. In particular, the component may transform from inactive into surface-active and vice versa. This conclusion is confirmed by the results of calculations for all investigated ternary and quaternary systems.

It is well known that the symbaticity criterion proposed by Rusanov, being the analogue of the Konovalov–Vrevskii law, is completely fulfilled only in binary systems. Therefore, without making any generalisations, it is noted that with the variation of composition in the beam sections in the investigated multicomponent systems, the composition of the surface layer changes in a symbatic fashion with the composition of the solution.

Figures 6.16–6.18 show the dependences of the molar surface and thickness of the surface layer on the lead content of the melt and in the surface layer. The isotherms of these parameters in the investigated ternary systems described by smooth curves without extrema. The relief of the isoteric surfaces of the discussed parameters is not simple, as indicated by the isotherms of the molar surface in the indium–tin–gallium ternary system (Fig. 6.18).

If the isotherm of the molar surface in the section $x_{In}:x_{Sn}=1:9$ is

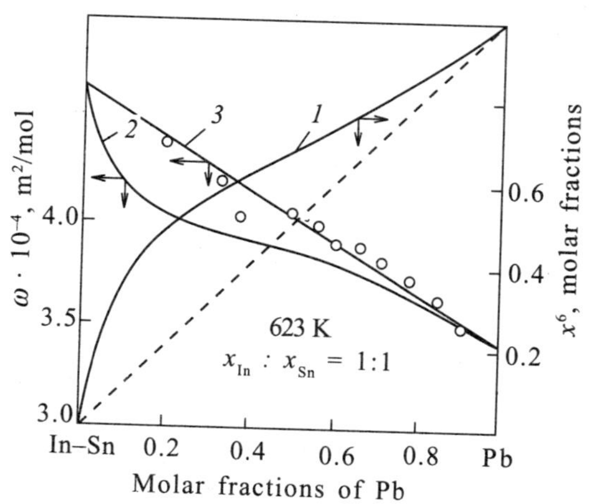

Figure 6.16. Isotherms of the molar surface and molar fraction of lead in the surface layer of indium–tin–lead melts at 623 K. 1 – the lead content of the surface layer; 2, 3 – the dependence of the molar surface on the lead content in the volume and the surface layer, respectively.

characterised by positive deviations from the additive straight line, then with the replacement of indium in the section with tin, the value of these deviations decreases (In–Sn=1:1) and at In:Sn=9:1 becomes negative.

These special features of the concentration dependence of the molar surface are caused only by the difference of the composition of the surface layer in relation to the composition of the volume phase. This is confirmed by the additivity of the molar surfaces in relation to the composition of the surface layer. The calculations show that in the investigated ternary and quaternary systems, the dependence of the molar surfaces on the composition of the surface layer is expressed by the additive function. Consequently, the partial–molar surfaces of the component are equal to their molar surfaces. In chapter 5, we derived an improved equation for calculating the partial–molar surfaces of the component from the volume properties of the melt. The calculations carried out using this equation confirmed the results.

Thus, the calculations of the partial–molar surfaces, performed by two independent methods–on the basis of the properties of the volume phase (molar and partial–molar volumes) and on the basis of the concentration dependence of the surface tension, gave the same result. The results of these calculations show that in the investigated melts, the partial–molar surface of the component is equal to its molar surface. This result greatly simplifies the calculation of surface tension using the well-known theoretical equations, in particular, the Popel'–Pavlov equation.

Figure 6.17 shows the dependence of the thickness of the surface layer of the indium–tin–lead ternary melts on the lead content in the

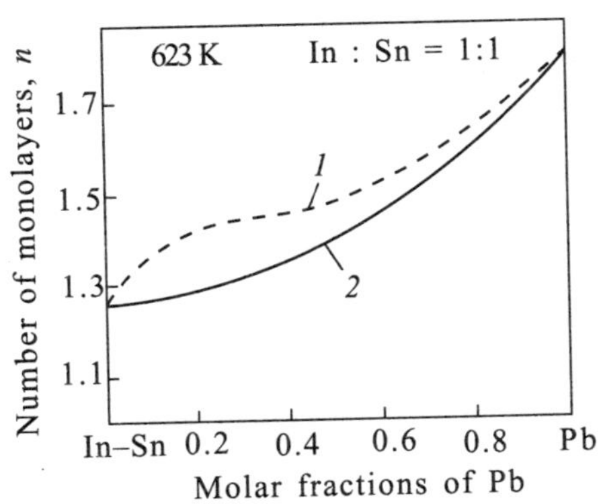

Figure 6.17. Isotherms of the thickness of the surface layer of indium–tin–lead melts at 623 K. 1, 2 – dependence of the thickness on the molar fractions of lead in the volume and in the surface layer of the melts.

volume (curve 1) and in the surface layer (curve 2). It should be mentioned that the behaviour of the isotherms of the thickness of the surface layer in the remaining systems is the same and, consequently, the graphs for these systems are not presented here. The values obtained for the thickness of the surface layer of pure metals indium, tin and lead in chapter 4 were: 1.5; 1.0; 1.8 atomic layers.

These values were used in calculations for the ternary system. As shown in Fig. 6.17, the thickness of the surface layer of the ternary solutions increases with increasing lead concentration. The rapid increase of the concentration is replaced at 0.2 molar fractions by a flat section with a small change of thickness. A further increase of the lead content results in a monotonic increase of the thickness of the values for pure lead. Identical isotherms of the thickness of the surface layer were also obtained for the indium–lead and tin–lead lateral binary systems.

Analysis of the curves shows that as in the case of the molar area, the variation of the thickness of the surface layer may be explained by sorption processes in the melts. In fact, the comparatively rapid increase of the thickness in the initial section of the isotherms is caused by the positive absorption of lead. In the

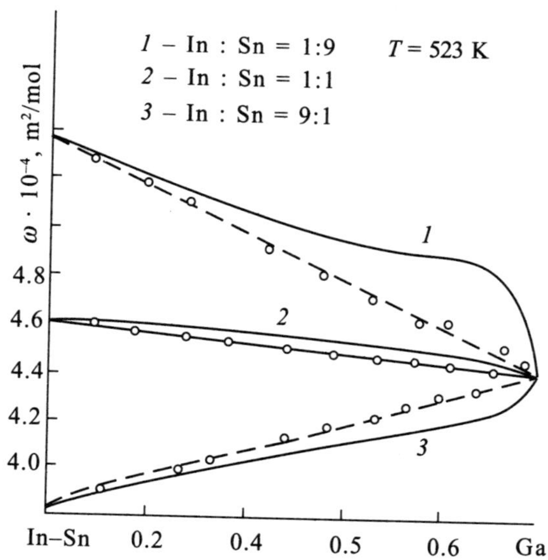

Figure 6.18. Isotherms of the molar surface of the indium–tin–gallium melts at 523 K. Dependence of the molar surface on the gallium content in the volume (solid line) and in the surface layer (dashed line).

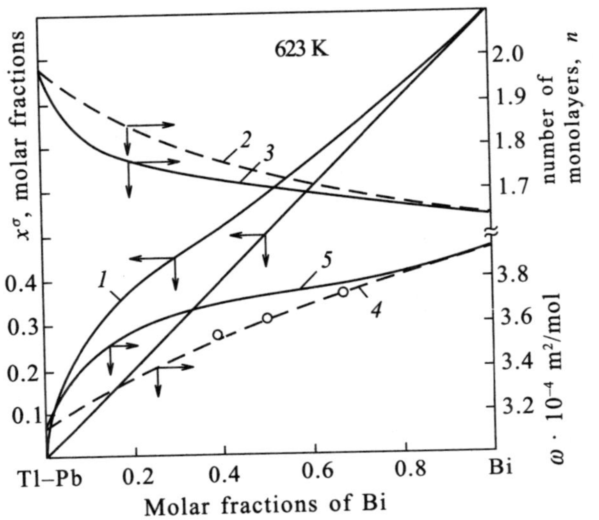

Figure 6.19. The isotherms of the parameters of the surface layer of thallium–lead–bismuth melts. 1–the molar fractions of bismuth in the surface layer; 2, 3 and 4, 5 – dependence of the thickness of the surface layer and the molar surface on the bismuth content of the surface layer and in the volume of the melts, respectively.

dilute solutions, as a result of adsorption, the lead content in the surface layer increases at a higher rate than in the volume of the solution. When the thickness of the surface layer of lead is greater than that of indium and tin, the thickness increases rapidly in this section of the isotherms. With the saturation of the surface with the lead atoms, the rate of the process decreases and a further increase of the concentration of lead results in a monotonic increase of the thickness of the surface layer up to the value for pure lead. The adsorption nature of this phenomenon is confirmed by smooth curves, characterising the dependence of the effective thickness on the lead content of the surface layer.

The relationships governing the variation of the parameters of the surface layer (composition, thickness, molar and partial–molar surfaces) in ternary systems are also characteristic of the four-component systems. This is also indicated by the isotherms of these properties, shown in Figs. 6.20 and 6.21. To avoid repeating ourselves, we restrict comment to the fact that the isotherms of the surface properties in the beam sections of the investigated ternary and quaternary systems do not show any special features not predicted by their isotherms in the appropriate lateral binary systems.

If the dependence of the molar surface of the bismuth content in the surface layer (curve 2, Fig. 6.20) is expressed by the additive line, then this value is almost completely independent of the composition of the solution in the beam section In–Sn–Pb = 1:1:1. This is due to the fact that the value of the molar surface of the

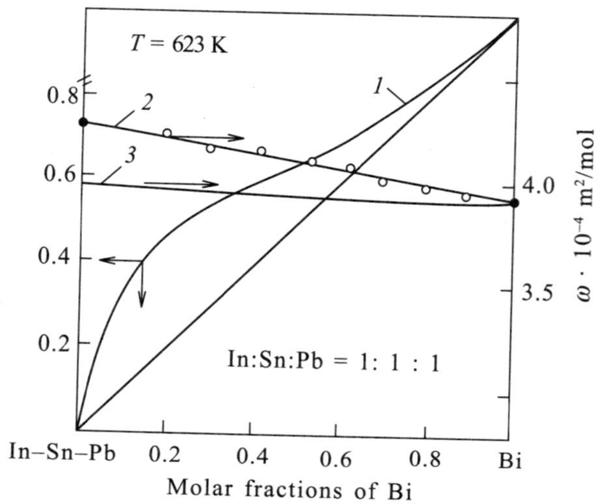

Figure 6.20. The isotherms of the molar surface and bismuth content of the surface layer of indium–tin–lead–bismuth melts at 623 K. 1 – the isothermal molar fractions of bismuth in the surface layer; 2, 3 – the dependence of the molar surface on the bismuth content in the surface layer and in the volume of the melts, respectively.

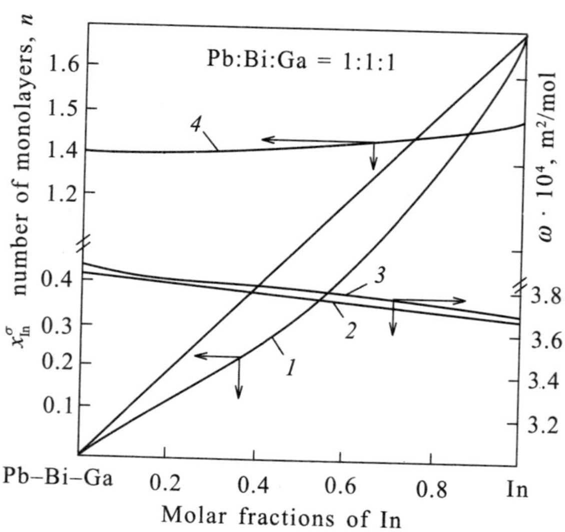

Figure 6.21. The isotherms of the parameters of the surface layer of the indium–lead–bismuth–gallium melts at 1073 K. 1 – the molar fraction of indium in the surface layer; 2, 3 – the dependence of the molar surface on the molar fractions of indium in the surface layer and in the volume of the melt, respectively.

In–Sn–Pb ternary alloy with the equimolar composition slightly differs from the molar surface of bismuth. A similar situation is also found in the case of the isotherms of the thickness of the surface layer.

6.3. The current state of experimental examination of the surface properties and molar volumes of multicomponent metallic systems

Analysis of our experimental data for the surface tension and molar volumes of the ternary and quaternary systems shows that the special features of the concentration dependence of these properties in the multicomponent melts are caused by their isotherms in the lateral binary systems.

It is interesting to verify the validity of this conclusion on other, experimentally investigated systems.

As mentioned previously, at the moment of formulation and definition of the current task, the surface tension of only nine ternary metallic systems had been investigated. In the last 15 years, this number greatly increased and, including our investigations, approximately 40 ternary and two quaternary melts have been studied. However, the results are scattered in different sources and have not been subjected to critical analysis and theoretical generalisation.

Therefore, in this section, it is attempted to examine critically the experimental data available in the literature on the surface tension and molar volumes of the multicomponent metallic melts. The experimental data were analysed to determine a relationship between the special features of the concentration dependence of the surface tension of multicomponent melts and lateral binary systems.

Only the studies concerned with the examination of the interface between the multicomponent metallic melt and the saturated vapour in the equilibrium condition will be considered. Attention is given to the results obtained in knowingly satisfactory conditions or using commercial purity metals. We do not take into account the studies concerned with the measurement of the surface tension of individual alloys for commercial applications.

The examination of the surface tension and density of the multicomponent systems in the entire range of the compositions started relatively recently. The main part of these investigations was carried out in the following sentence: the Ural Polytechnical Institute (Ekaterinburg), the Chechen-Ingush State University (Groznyi), the Kabardino-Balkarsk State University, the Institute of Problems of Materials Science (Kiev) and the Moscow Institute of Steel and Alloys.

The concentration dependence of the surface tension of Hg–Cd–K and Hg–Cd–Cs ternary systems in a narrow composition range was investigated for the first time in [253, 254]. Measurements of the surface tension of 266 ternary solutions were taken at the same temperature of 295 K. The composition of the melt was varied from 0 to 0.070, 0.0054; 0.000025 molar fractions of cadmium, potassium and caesium, respectively.

In the investigated range of the compositions, the authors detected several special features of the variation of surface tension with composition.

Experimental investigation showed that at molar fractions of alkali metals of $x_k = 0.004$ and $x_k = 5 \cdot 10^{-7}$, surface tension is independent of the cadmium content, *i.e.*, the phenomenon of concentration buffering was detected. It was shown previously that this phenomenon is also typical of ideal multicomponent systems and may be predicted on the basis of the isotherms of surface tension of lateral binary systems.

In [256, 257], the surface properties of iron–cobalt–nickel and iron–manganese–silicon melts were studied at 1723 K. Measurements were taken by the 'large' droplet method on a substrate of aluminium oxide at the interface with an inert gas. All the isotherms of σ of the Fe–Mn–Si melts are almost completely linear, *i.e.*, the surface tension isotherms of both binary and of sections of the ternary system are described by the linear functions of the molar fractions of the components. The isothermal surfaces of the surface tension of this system are described fully by the Popel'–Pavlov equation for ideal solutions so that it was possible to carry out strict calculations of the absorption of components in the investigated system. The results of the calculation showed that iron is adsorbed positively in the entire range of the compositions, and cobalt negatively. Nickel in the ternary alloys adsorbs both positively and negatively, depending on the composition of the melt.

In the Fe–Mn–Si ternary system [257] the surface tension isotherms in the sections corresponding to the constant content of iron and manganese, like the isotherms of the Mn–Si binary system, show the presence of maxima. With the increase of the iron content, the position of the maximum is shifted slightly in the direction of the Fe–Mn axis. The surface tension isotherms in the Fe–Si binary system at 0.50 molar fraction of iron are characterised by points close to angular. At the same iron concentration, the points, close to angular, are detected in the sections with a constant ratio of the atomic fractions of manganese and silicon. As the section approaches the Fe–Mn axis, the form of the isotherms becomes similar to the isotherms of the binary system.

The surface tension and density of the melts of the Sn–Pb–Bi, Sn–Ga–Zn and Fe–Cr–Mn systems was examined by the authors of [258, 271]. The surface tension of the first two systems was determined by the method of maximum pressure in the droplet, and the density with a pycnometer. The surface tension of the Fe–Cr–Mn melts was measured by the sessile droplet method [271].

The isothermal surfaces of the molar volumes and surface tension of the Sn–Pb–Bi ternary systems are smooth, without singular points.

The shape of the isothermal surfaces of σ of the Sn–Ga–Zn system is more complicated [258]. The isotherms of σ of the gallium–tin system are smooth, are the surface tension isotherms of the Sn–Zn and Ga–Zn systems show a minimum in the range of the low zinc content. It should be mentioned that the same shape of the isotherms was recorded in the case of the variation of composition in the sections with a constant ratio of the molar fractions of tin to gallium. The molar volumes of both binary and ternary melts are characterised by small positive deviations from additivity.

The isothermal surfaces of surface tension in the Fe–Cr–Mn ternary system are smooth, without singular points, and are in agreement with the smooth isotherms of σ of the boundary binary systems. The isotherms of the molar volumes in the investigated composition range may be described by an additive function [258].

The smooth isothermal surfaces of σ and V in the In–Pb–Bi systems were constructed by the authors of [259]. It has been established that when bismuth is replaced with lead the surface tension isotherms and the isotherms of the molar volumes of the indium–bismuth binary system gradually transform to the isotherms of these properties in the indium–lead system.

The results of experimental measurements of the surface tension and density of the lead–bismuth–mercury ternary system are interesting [260]. This is one of the few ternary metallic systems for which experimental data on the thermodynamic activity coefficients are available [261]. This enabled the authors to calculate many properties of the surface layer of the ternary melts and explain the results on the basis of considerations regarding the special features of the density of the energy states of mercury atoms.

A characteristic feature of the isothermal surfaces of surface tension of the investigated system is the presence of a fold forming maxima in binary amalgam systems. In the lead–mercury and bismuth–mercury binary systems the isotherms of σ showed maxima (in the range of mean compositions in the Pb–Hg systems and closer to the pure mercury in the Bi–Hg system). It should be stressed that identical maxima have been detected on the isotherms of σ of all investigated sections with a constant lead to bismuth ratio. In this case, in replacement of bismuth with lead, the maximum is shifted from the region of the compositions, enriched with mercury, to the region of the mean compositions. The densities of the Hg–Bi–Pb alloys were used to calculate molar volumes. The isothermal surfaces of the molar volumes are smooth, without singular regions.

The authors of [262, 263] used the sessile droplet method for measurement of σ and ρ of mutually saturated solutions of Fe–Ni–Ag, Co–Ni–Ag [261] and Fe–Cu–Ag, Fe–Si–Ag, Fe–Mn–Ag, Fe–Al–Ag ternary systems [263]. The results show that the surface tension of the Ag–Me–Fe melts differs only slightly from that of the Ag–Me alloys. The surface tension of the ternary Fe–Me–Ag systems decreases more rapidly than the surface tension of the Fe–Me binary systems. This is determined by the increase of the activity of sulphur $(\partial\sigma/\partial x_{Hg})_{x_{Hg}}$ in the presence of a third component.

The temperature dependence of the surface tension and molar volumes of the Fe–Co–Si and Fe–Ni–Si melts, distributed in the sections with a constant silicon content, has been examined in [264, 294]. All the melts were characterised by negative deviations of the molar volumes from additivity. In both the Fe–Si and Co–Si binary systems and in ternary systems, the maximum deviations were found at 0.5 molar fractions of silicon.

It is interesting to note that the partial–molar volumes in the sections with the constant ratio of iron to cobalt show significant changes at the same compositions as in the binary melts. The molar volumes of the ternary melts in the sections with the same silicon content are similar to the mean values for the Fe–Si and Co–Si binary systems [293]. The isotherms of the temperature coefficient of surface tension are characterised by a low maximum at 0.1–0.2

and by a minimum at 0.3–0.4 molar fractions of silicon. Similar extrema have also been detected at the same concentration of silicon in the Fe–Si and Co–Si binary systems.

The results of the examination of the density surface tension of the Pt–Pd–Rh and Pb–Sn–Ag melts are presented in [265, 267] respectively. Examination of the Pb–Sn–Ag system shows that the isotherms of σ of the ternary alloys, containing up to 0.2 molar fraction of silver, reproduce almost completely the appropriate isotherm of the Pb–Sn binary system. According to the authors, this is caused by the fact that the silver content in the surface layer of both the ternary and binary melts is close to zero.

In [265, 266] it was shown that the equation of zero approximation of the theory of regular solution and the Zhukovitskii equation satisfactorily describe the isotherms of σ and V of the Pt–Pd–Rh, Pt–Pd–Si, Pt–Rh–Si and Rh–Pd–Si ternary systems. Consequently, the authors concluded that these equations may be used to calculate the values of σ of ternary systems with the unlimited mutual solubility of the components.

The isotherms of the molar volumes of the ternary systems investigated in [289] show negative deviations from the additivity and considerable asymmetry. Surface tension changes non-monotonically. There are inflection regions of the isotherms of σ. However, according to the authors of [289], the pattern detected in the ternary melts differs only slightly from the results obtained for binary systems with silicon [290–294].

The smooth isothermal surfaces of surface tension and specific volumes were produced by the authors of [273, 274, 278] in seven ternary systems: Ni–Mn–Ge, Ni–Mn–Cu, Ni–Cr–Ge, Ni–Mn–Sn, Ni–Cr–Sn, Ni–Cr–Cu, Ni–Mn–Ga. If the formation of Ni–Cr–Sn, Ni–Cr–Ge, Ni–Mn–Ge and Ni–Mn–Sn alloys is accompanied by a decrease of volume, a decompression is found in the Ni–Cr–Cu system. The isotherms of the specific volumes of the Ni–Mn–Cu system differ only slightly from the additivity, indicating the absence of any compression in the melt.

The experimental isotherms of σ in the sections of the ternary systems are situated below the Zhukovitskii ideal isotherms. No 'ternary' effects were detected on the isotherms of the investigated properties. The comparison of the isotherms in the triple sections with the isotherms of the lateral binary sections is difficult because of the absence of reliable data on the binary systems. In particular, there are no data on the isotherms of σ of the Mn–Sn binary system in a wide composition range.

The surface tension of the copper–tin–silver ternary system was investigated in [277]. It is shown that the isotherms of σ in the sections are characterised by smooth curves which gradually transfer from the isotherms of the Ag–Pb melts to the isotherms of Cu–Pb when silver is replaced with copper.

The authors of [281] found an increase of the surface activity of rare-earth metals on aluminium in the presence of gallium. However, the absence of the data on the surface tension in the entire range of the composition does not make it possible to determine the relationship of the surface tension isotherms of the ternary and lateral binary systems.

The variation of the molar volumes and surface tension with composition in the Ga–In–Sn systems has been examined in [268, 269, 270]. The results show large deviations of the molar volumes from additivity in the range of the mean compositions of the ternary system and, consequently, the authors of [269, 270] concluded that the interaction between the atoms in the ternary alloy is intensified. The mixing enthalpy of the system was investigated in [271] by quantitative thermography. It has been shown that the mixing enthalpy of both binary and ternary melts is a positive value but this is not in agreement with the high negative deviations of the molar volumes.

Thus, of the investigated ternary systems, only these melts showed the so-called ternary effects not predicted by the isotherms of the lateral binary systems. For this reason, the temperature dependence of σ and ρ of the melts of these system was again investigated by us [272]. The results were discussed in chapter 5 where it is shown that the isotherms of σ in the sections are smooth, without extreme points, and the isotherms of the molar volumes are close to additive straight lines.

The Kabardino-Balkarsk Institute has carried out in recent years, under the supervision of Kh.B. Kokonov and B.A. Alchagirov, systematic investigations into the surface properties and molar volumes of the binary and ternary systems based on alkali metals [113, 114]. New results were obtained in this work are of considerable scientific interest.

In particular, the authors investigated the surface tension and density of sodium–caesium–potassium [112] and sodium–rubidium–caesium ternary systems [113]. On the basis of the results we have concluded that the concentration dependence of the surface tension of the ternary systems may be predicted on the basis of the surface tension isotherms of the lateral binary systems. The experimental results show that in the systems the surface tension isotherms into being sections of the concentration triangle have the form of smooth curves, without singular points, which is in agreement with the identical curves in the lateral binary systems.

Thus, the results of our investigations and analysis of the data show that the concentration dependence of the surface tension and molar volumes of the multicomponent melts do not show special features which are detected to some extent on the isotherms of these properties of the lateral binary system. The special features of the isotherms of the surface tension of the lateral binary systems are manifested on the isothermal surfaces of ternary melts. An increase of the content of the third component results in smoothing of the special features. The intensity of this process depends directly on the magnitude and sign of the limiting surface activity of the third component.

Therefore, the experimental data published in the literature confirms the conclusion made in chapter 5 according to which the form of the isothermal surfaces σ and ρ in the ternary metallic melts is determined by the concentration dependence of these properties in the lateral binary systems. This indicate that it may be possible to predict the properties of multicomponent melts on the basis of their isotherms in binary system.

6.4. Methods of predicting the surface properties of multicomponent melts

The experimental examination of the surface properties of multicomponent melts is associated with considerable difficulties, and the calculations carried out using the currently available theoretical equations either require the data on other properties of these melts or do not describe the surface tension isotherms. In these conditions, the role of simple and reliable semi–empirical methods of evaluating σ of the multicomponent systems becomes more important. Work been carried out for a long period of time to develop these methods for the volume properties, such as molar volumes, viscosity, excess thermodynamic potentials, and also for plotting equilibrium diagrams [252, 282, 283].

Recently, attempts have been made to predict the adsorption properties of solids [284–287]. As far as we know, there are no studies in the literature concerned with the development of similar methods of predicting the surface properties of the melts, although the importance of these methods is obvious. Therefore, we have attempted to link the surface tension of a multicomponent solution with the properties of lateral binary systems. Several methods have been proposed and the optimum method selected [288]. These methods will be examined in greater detail.

The prediction methods, based on the Zhukovitskii equation

The principle of all semi-empirical methods of prediction is that, describing by some function the dependence of σ in the composition in the boundary binary systems, graphical or analytical interpretation is used to determine the values of surface tension for multicomponent alloys, in the simplest case by defining in advance the dependence (σ) of the multicomponent systems on the composition in the form of a polynomial of the n-th degree, and the polynomial coefficients can be determined from the experimental values of the surface tension of the appropriate binary systems. However, this leads to formally mathematical methods. Therefore, when developing prediction methods, our work was based on the well-known physical equations, in particular, the Zhukovitskii equation. The coefficients of this equation have a specific physical meaning. On the other hand, in transition from the binary to multicomponent systems, the number of terms in the equation does not increase but the value of the activities changes in both the surface layer and the volume phase.

Thus, using the Zhukovitskii equation for predicting the surface tension of multicomponent solutions reduces the problem to the determination of the thermodynamic activities of the components in the volume phase α_i and in the surface layer α_i^{ω} of the multicomponent solution. If these values are available, it is possible to calculate the surface tension of the multicomponent system using the equation:

$$\sigma = \sigma_i - \frac{RT}{\omega_i} \ln \frac{a_i^{\omega}}{a_i} \qquad (6.4.1)$$

where σ_i and ω_i are the surface tension and molar area of the pure i-th component, respectively. The values of α_i in the multicomponent system, if they are not determined by experiments, can be determined by one of the currently available methods [252, 261] using the data on the values of the thermodynamic activities of the components in the lateral binary systems. The thermodynamic activities of the components in the surface layer of the lateral systems are calculated from the Zhukovitskii equation. The experimental data for σ of these alloys were used in these cases. From the formally mathematical viewpoint, α_i^ω is a fitting parameter. The values of this parameter for the lateral binary systems are selected to ensure that the equation (6.1) describes the experimental isotherm of surface tension.

Having the values of α_i^ω for the lateral binary systems and examining the surface layer as a separate phase, we can calculate the thermodynamic activity of the component in the surface layer of the multicomponent melt. For this purpose, as in the case of the volume phase, we use the previously mentioned methods [261], Table 23.

After calculating α_i and α_i^ω from equation (6.1), it is possible to calculate the surface tension of the multicomponent melt. However, in this case, calculation of σ is associated with difficulties due to the fact that α_i^ω is defined as the function of the composition of the surface layer of the multicomponent solution. Therefore, to calculate σ for the given composition, it is necessary to determine the content of the component in the surface layer using the approximate methods. Undoubtedly, this complicates the calculations and reduces the accuracy of prediction.

The difficulties, associated with the need to evaluate the composition of the surface layer, may be avoided if α_i^ω is treated as the function of the composition of the solution and it is assumed that the semiempirical methods of calculating the thermodynamic activity may also be used for this relationship. However, this assumption is permissible in the case in which the solutions of both the binary and multicomponent system are close to ideal. On the other hand, the behaviour of the i-th component in the surface layer is also important. If the i-th component (whose thermodynamic activity in the surface layer α_i^ω is calculated) shows a strong capillary activity in the solutions, the content of the component in the surface layer is considerably higher than in the volume phase and this results in a large reduction of prediction accuracy. Consequently, the i-th component should be represented by the component with the lowest absolute surface activity.

We carried out calculations, using both methods, for the indium–tin–lead melts. The initial component was either indium or tin, and the calculations show that the results, obtained by the first methods, are in better agreement with the experimental data than the data calculated using the second method. The maximum difference is 1% in the first method and 3% in the second method.

For the Tl–Pb–Bi system, calculations of σ were carried out using the first method. Comparison of the results with the experimental data shows that the maximum difference is 2.5%.

6.5. Prediction of surface tension and excess molar volume on the basis of the properties of melts of lateral binary systems

As mentioned previously, the Koller method is one of the optimum method of predicting the excess properties, in particular, the excess molar volume [252]. The application of this method for predicting surface tension is associated with difficulties because excess quantities are present in the final formula. Therefore, we have introduced conventionally the concept of the excess surface tension:

$$\Delta\sigma = \sigma - \sigma_{id} \tag{6.5.1}$$

where σ is the surface tension of the solution; σ_{id} is the surface tension of the ideal solution.

Since there are a number of various models of the ideal solution, it should be mentioned that any equation, presented in the literature, may be used for the dependence of σ_{id} on composition. The main requirements of the function: simple mathematical procedure and absence in the equation of difficult to determine or unknown physical quantities. In most cases, we used the well-known expression [247]

$$\sigma_{id} = \sigma_j - \frac{RT}{\omega} \ln \sum_{i=1}^{n} x_i \exp\left[\frac{\sigma_i - \sigma_j}{RT}\right] \tag{6.5.2}$$

where σ_i and σ_j are the surface tension values of the pure component; $\omega = \frac{1}{n}\sum_{i=1}^{n}\omega_i$ is the molar area averaged out with respect to all components; x_i is the molar fraction of the i-th component.

Further calculation showed that the accuracy of prediction does not decrease if the equation (6.5.2) is replaced by the additive function:

$$\sigma = \sum_{i=1}^{n} \sigma_i x_i \tag{6.5.3}$$

It should be mentioned that regardless of the comparatively smaller 'physical' meaning, the application of the additive function (6.5.3) greatly simplifies the calculations without impairing the accuracy of prediction.

We examine in greater detail the derivation of the final formula. This will show the strong and weak aspects of the proposed prediction method. As carried out by Koller for the volume properties, the dependence of the excess surface tension on composition in the lateral binary systems is represented in the form:

$$\Delta\sigma_{ij} = f\left(K_{ij}\right)x_i'x_j' \tag{6.5.4}$$

Here x_i' and x_j' are the molar fractions of components in the binary system i–j; $K_{ij} = \frac{x_i}{x_j}$ is the ratio of the number of the moles of the component; $f(K_{ij})$ is a function which depends on the composition of the binary alloy.

It is assumed that the following relationship holds for the multicomponent alloy:

$$\Delta\sigma = \frac{1}{2}\sum_{i,j=1}^{n} f\left(K_{ij}\right)x_i x_j \tag{6.5.5}$$

where x_i and x_j are the molar fractions of the components in the multicomponent melt. Equation (6.6) holds in the case in which an increase of the content of the third component reduces the contribution brought in by the interaction of different atoms, in accordance with the additive law.

The expression below the sum sign in (6.5.5) is transformed

$$f\left(K_{ij}\right)x_i x_j = f\left(K_{ij}\right)\frac{N_i}{N}\frac{N_j}{N} = f\left(K_{ij}\right)\frac{N_i}{N_i+N_j}\frac{N_j}{N_i+N_j}\left(\frac{N_i+N_j}{N}\right)^2$$
$$= f\left(K_{ij}\right)x_i'x_j'\left(x_i+x_j\right)^2 = \Delta\sigma_{ij}\left(x_i+x_j\right)^2 \tag{6.5.6}$$

where N_i and N_j is the number of moles of the i-th and j-th components, and $N = \sum_{i=1}^{n} N_i$ is the total number of moles of all components in the solution.

Thus, for the excess surface tension of the multicomponent solution:

$$\Delta\sigma = \frac{1}{2}\sum_{i,j=1}^{n}\Delta\sigma_{ij}\left(x_i+x_j\right)^2 \tag{6.5.7}$$

where $\Delta\sigma_{ij}$ is the excess surface tension of the two–component solution *i–j* with the same ratio of the molar components as in the multicomponent solution. The determination of $\Delta\sigma_{ij}$ on the basis of the experimental isotherms of the surface tension of binary systems is associated with certain difficulties. Therefore, this problem will be examined in greater detail.

Initially, using the well-known expression $K_{ij} = \frac{x_i}{x_j}$, for the multicomponent solution, from the equation

$$K_{ij} = K_{ij}^{'} = \frac{x_i'}{1 - x_i'} \tag{6.5.8}$$

we determine the appropriate composition of the binary alloy x_i'. The primes indicate the quantities relating to the binary systems. Knowing the composition of the binary alloy, the surface tension is calculated from (6.5.2) or (6.5.3). Subsequently, the excess surface tension $\Delta\sigma = \sigma - \sigma_{id}$ is calculated from the experimental isotherms of the surface tension of the *i–j* binary system for the given composition x_i'. Determining by the same procedure the values of $\Delta\sigma_{ij}$ for other binary systems, from equation (6.8) we can easily determine $\Delta\sigma$ and, consequently, the surface tension of the multicomponent solution.

Thus, the expression (6.5.7) may be used to determine the surface tension of the multicomponent solution if the dependence of the surface tension on composition for all lateral binary system is available.

The excess molar volume is calculated directly from the Koller equation:

$$\Delta V = \frac{1}{2}\sum_{i,j=1}^{n} \Delta V_{ij}\left(x_i + x_j\right)^2 \tag{6.5.9}$$

The proposed equations were used to calculate the surface tension and molar volumes of 16 ternary and 2 quaternary systems for which the experimental data were available. The results are discussed below. Some of the experimentally investigated ternary systems could not be calculated because of the absence of data on the surface tension of the lateral binary systems.

Cases may occur in practice in which the measurement of the physical–chemical properties of a binary system is associated with difficulties determined by the high melting point, chemical 'activity', the risk of explosion, *etc*. In these cases it is desirable to have reliable methods of evaluating the properties of binary systems on the basis of the properties of ternary ones.

The examined method may also be used to solve this problem. In fact, if necessary, the proposed equation may be used for calculating the surface tension of one of the lateral binary systems using the data on σ of multicomponent alloys. For example, the experimental data on the surface tension of Ni–Mn–Sn melts at 1523 K were used to determine the concentration dependence of the surface tension of the Mn–Sn binary system. The surface tension of the Mn–Sn system at a low tin concentration was measured by experiments in [12]. Our results obtained for the given concentration range are in agreement with the experimental data.

On the basis of the shape of the isothermal surface of surface tension, the ternary melts may be divided into two groups.

The first group includes the systems with smooth isothermal surfaces of surface tension. It should be mentioned that the smooth isothermal surfaces of σ can form only in cases in which the isotherms of σ of all lateral binary systems are smooth curves without extrema.

The second group includes multicomponent systems characterised by a complicated dependence of surface tension on composition. In particular, these are melts with a minimum, a maximum or points close to angular, on the isotherms of the surface tension of one of the lateral binary systems.

Surface tension was calculated for the system of both the first group and for more complicated systems included in the second group. The results are presented in Figs. 6.22–6.26. The solid line indicates the experimental data, the dashed line the calculated values. The figures show ternary systems characterised by the largest deviations of the calculated values from the experimental data.

Comparison of the calculated isotherms of σ with the experimental values shows that if in the systems Hg–Pb–Bi, Fe–Mn–Si, belonging to the second group, the proposed equation describes only quantitatively the concentration dependence of the surface tension, then for all systems of the first group and for the majority of the systems of the second group, the calculated values coincide in the overall experiment error range with the experimental data, and in the In–Sn–Pb–Bi system the calculated values are in agreement with the experimental values in the experimental error range.

The Koller method was used for predicting the values of the

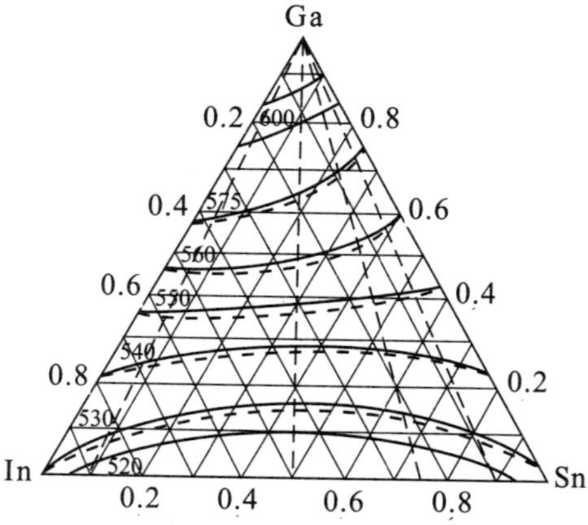

Figure 6.22. Surface tension of the indium–tin–gallium ternary system at 773 K. Solid line – experimental values, dashed line – calculated values.

molar volume of multicomponent melts. We calculated the excess molar values of more than 10 ternary systems for which the experimental data were available. The most interesting results are presented in the figures. In all the systems, the calculated isotherms of excess volume are in agreement with the experimental data.

Thus, the results of the calculations of the surface tension and molar volume show that the proposed method is suitable for predicting the properties of multicomponent notes. If the isotherms of σ of all lateral binary systems contain no extrema, the error of prediction does not exceed the total error of measurement of surface tension.

The practical importance of any method of the calculation is determined by two factors: its accuracy and availability of the initial data. The accuracy of the proposed method is sufficient (the error of prediction for the majority of ternary systems does not exceed the measurement error). The initial data are represented by the values of σ of the binary melts presented in the handbook literature.

The main shortcoming of the method is the absence of physical justification of the final formula, *i.e.*, the physical model of the solution, for which the final formula holds. Regardless of attempts made in this direction, the method may be regarded as mathematically formalised. The Koller's conclusion can be regarded as theoretically justified and clear. The main scientific value of the proposed method, as of any other similar method, is evidently the formation of prerequisites for the derivation of theoretical equations, based on the physically justified models of solutions.

The attempt for the application of the proposed method for

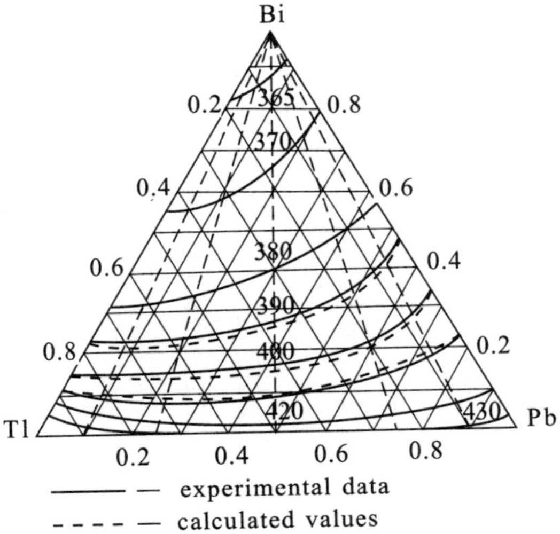

Figure 6.23. Surface tension isolines of thallium–lead–bismuth melts at 773 K.

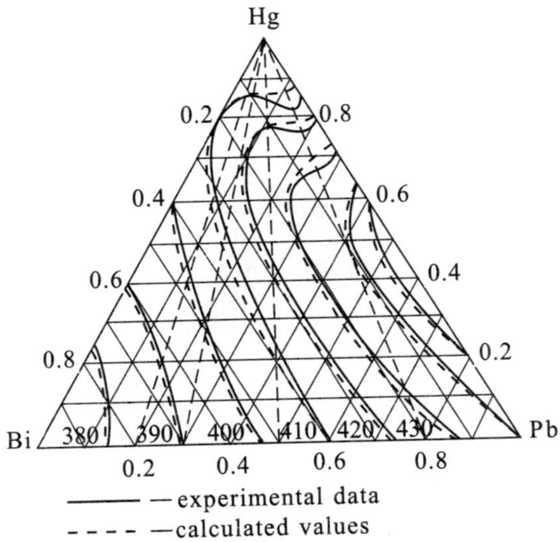

Figure 6.24. Surface tension of lead–bismuth–mercury melts at 623 K.

calculating the surface tension has proved to be highly efficient, and the area of application will be widened.

The vast amount of experimental material, available for the binary systems, and the limited number of the investigated multicomponent melts increase the practical importance of the method and will promote is further extensive application.

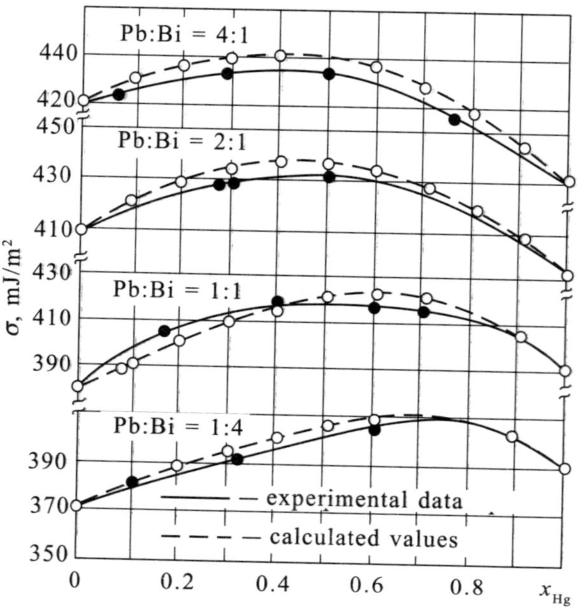

Figure 6.25. Surface tension isotherms of lead–bismuth–mercury melts at 623 K.

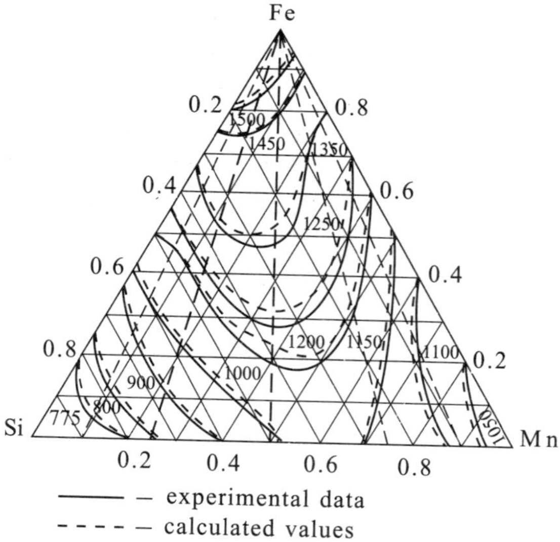

Figure 6.26. Surface tension of iron–manganese–silicon melts at 1823 K.

6.6. Theoretical description of the concentration dependence of the surface tension of multicomponent solutions

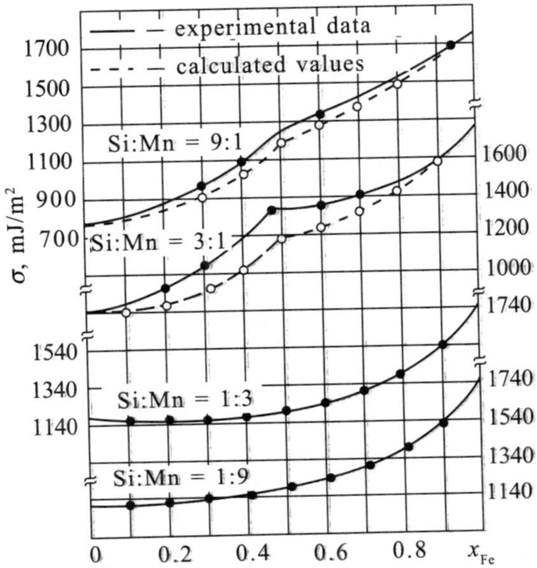

Figure 6.27. Surface tension isoytherms of iron–manganese–silicon melts at 1823 K.

As shown previously, the data on the dependence of the surface tension on composition and temperature may be used to find the thermodynamic parameters of the surface layer. Usually, this dependence is determined by expensive time-consuming experimental investigations. The labour content of the experiments increases greatly in the examination of the multicomponent metallic systems. Therefore, it is not surprising that special attention is given to the theoretical description of the surface properties of real multicomponent systems. The availability of justified equations, describing the dependence of surface tension on composition and temperature, would make it possible to find many thermodynamic parameters of the surface layer of the systems which have not been examined in the experiments and, consequently, predict the adsorption characteristics.

A large amount of experimental data is available on the physical–chemical and surface properties of pure metals and binary systems. Theoretical equations have been derived [7, 13, 82, 298, 299] for describing the concentration dependence of the surface tension of a large number of binary systems using the data on phase equilibrium.

However, insufficient attention has been given to the theoretical and experimental examination of the surface properties of multicomponent systems. There is no strict theory for the

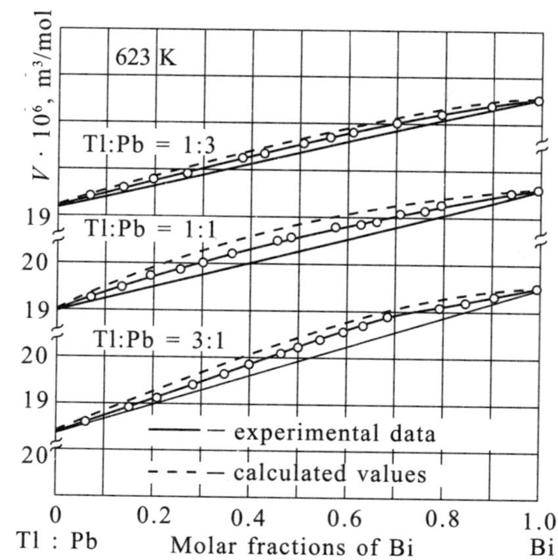

Figure 6.28. Isotherms of molar volumes in thallium–lead–bismuth melts at 623 K. Solid line – experimental data, dashed line – calculated values.

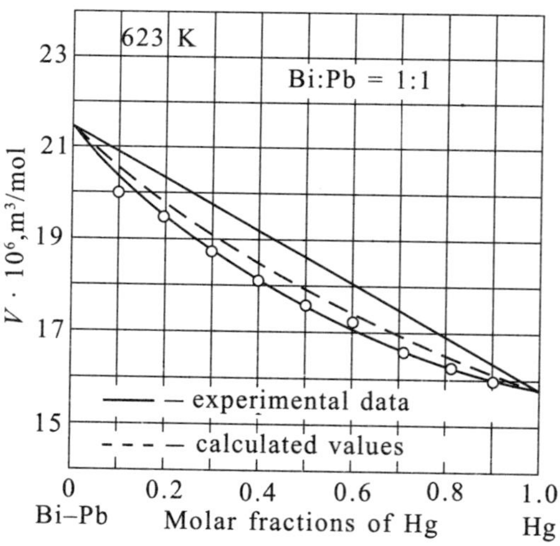

Figure 6.29. Isotherms of molar volumes in bismuth–lead–mercury melts at 623 K. Solid line – experimental data, dashed line – calculated values.

quantitative prediction of the dependence of surface tension on composition and temperature. The currently available theoretical equations either contain physical quantities was experimental determination is associated with insurmountable difficulties [82] or in many cases do not describe even quantitatively the surface

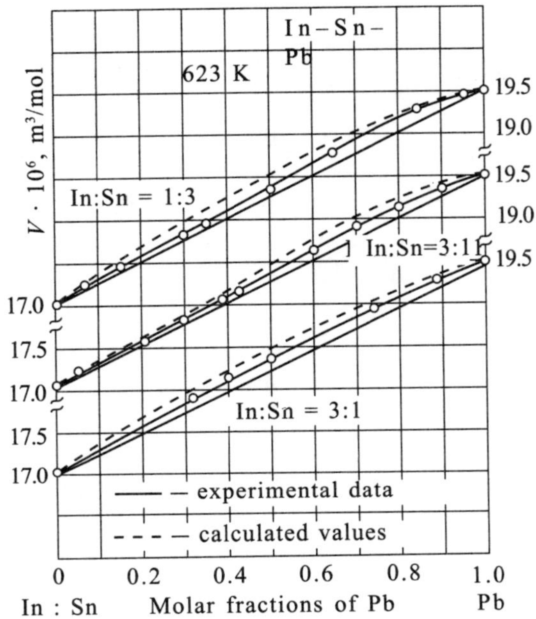

Figure 6.30. Isotherms of molar volumes in indium–tin–lead melts at 623 K.

properties of multicomponent melts. Therefore, the problem of finding stricter models for describing the surface properties of real multicomponent systems is still very urgent.

Analysis of the experimental data available in the literature on the surface tension of multicomponent systems, described in chapter 5, shows that special features of the concentration dependence of the surface properties of multicomponent systems are linked directly with the isotherms of these properties in lateral binary systems. Consequently, it is concluded that the concentration dependence of the dynamic parameters of the surface layer in multicomponent melts may be predicted on the basis of their isotherms in lateral binary systems. A semiempirical method of predicting the concentration dependence of the surface tension of multicomponent systems on the basis of the data on a similar dependence in lateral binary systems has been proposed. The main shortcoming of the proposed method is, as shown later, the absence of physical justification of the proposed formula. Therefore, it is important to derive an equation in which the surface tension of a multicomponent melt is linked with the properties of the lateral binary melts.

An increase of the number of component is accompanied by an increase in the difficulties in describing the nature of variation of the surface tension in relation to composition and temperature. In particular, this explains the incomparably smaller number of theoretical and experimental studies concerned with the examination of the surface tension of multicomponent systems.

The theoretical equations of the surface tension isotherms in a general form should describe a relationship of the surface tension with the concentrations of components and their physical–chemical properties. We believe that in multicomponent melts, these properties should also include the values of the surface tension of lateral binary systems. The solution of this problem in this general formulation is very difficult. Therefore, in deriving the equations it is necessary to make certain assumptions and allowances leading to different expressions for estimating the surface tension.

The currently available thermodynamic equations of surface tension isotherms may be divided into two groups. The first group includes the dynamic equations, based on the condition of equilibrium of the surface layer and the volume phase (2.4.2). These equations take into account the variation of the thermodynamic activities and molar surfaces with composition and temperature.

In [81] a more general relationship was described which differs from (2.4.2) by the fact that it takes into account the existence of many layers of the surface zone and its compressibility. Using this equation, Zadukmin and Khokonov [82] derived a general equation of surface tension isotherms of multicomponent systems. The resultant equation which is of a general nature, cannot be used for calculating surface tension in practice because data on the physical quantities are required here and the experimental determination of the data is faced with insurmountable difficulties. However, this equation, like partial cases, includes almost all known thermodynamic equations of the surface tension isotherms of the solutions. Transferring from the general to partial equations, this makes it possible to explain the assumptions in which the given equation is derived.

The second group includes the equations of surface tension isotherms. The derivation of these equations is based on the Gibbs adsorption equation. Usually, the adsorption equation is used for the determination of the adsorption and composition of the surface layer on the surface tension isotherms. However, this equation may also be used to solve an inverse problem. In fact, if the concentration dependence of adsorption is defined by some method, the integration of the Gibbs adsorption equation gives the equation of surface tension isotherms.

In the first chapter, we derive the equations for calculating various variants of adsorption in the multicomponent melts. This equation may be used to solve the inverse problem, *i.e.*, derivation of the equation for the concentration dependence of the surface tension of multicomponent solutions. A similar equation was derived relatively recently in [300, 301]. The authors of these studies obtained, for ideal binary and ternary solutions, an equation for the concentration dependence of surface tension from the Gibbs adsorption equation. To integrate the Gibbs adsorption equation, it is necessary to know the dependence of adsorption on the composition of the volume phase. This dependence was determined using the approximate equation published in Rusanov's book:

$$\frac{x_2^{\omega}}{x_1^{\omega}} = m\frac{x_2}{x_1} \tag{6.6.1}$$

The value of adsorption is associated with the molar fractions of the component in the surface layer through the molar area. Consequently, in integration, in addition to the expression (6.10), it is necessary to have the dependence of the molar area on composition. The author [300, 301] assumed that the molar area is an additive function of the composition of the volume phase. These assumptions greatly reduce the range of applicability of the resultant equations even for binary systems. The solution of this problem for multicomponent systems is made far more complicated.

Evidently, the problem of applicability of the equation of type (6.6.1) requires special examination, even for ideal ternary systems. This was not carried out by the author of [301]. We believe that these reasons may be used to explain the large differences in the values of the surface tension of indium–tin–lead ternary melts, obtained by the author of [301], in comparison with our experimental data [243]. The theoretical data, obtained in our study, are in almost complete agreement with the additive values.

In contrast to the binary systems, the composition of the n-component system is characterised by the n–1 variables (molar fractions) which can be changed irrespective of each other. The multicomponent systems contain an infinite number of the methods of changing the composition, whereas only one method is possible in the binary systems. Therefore, when discussing the problem of the validity of the thermodynamic equations in the multicomponent systems, it is necessary to use an approach which differs from that used in the case of the binary systems. In particular, when solving this problem, it is important to indicate the method of varying the composition of the multicomponent solution.

The problem of the validity of the Konovalov–Vrevskii laws in the multicomponent systems has been examined in some detail by Storonkin in [302]. A detailed analysis of these laws in different methods of changing the composition is provided. It is shown in which cases these laws are fulfilled in multicomponent systems. A detailed analysis will also be carried out in respect to other thermodynamic laws and equations. They include (6.6.1), Rusanov symbatic criterion, *etc.*

Therefore, we have attempted to derive an equation for the surface tension isotherms of multicomponent systems taking into account the circumstances mentioned above. We have used expressions for the adsorption and composition of the surface layer described in the previous chapters and this has greatly simplified the solution of the given task. As shown later, regardless of the assumption made regarding the ideal state of the solutions, the resultant equation describes satisfactorily the surface tension isotherms of a large number of binary and ternary systems and contains only the thermodynamic characteristics which can be determined by experiments with sufficient accuracy.

6.7. Equation of surface tension isotherms of binary and ternary systems

Taking into account the fundamental thermodynamic equations, we have derived the following equation for calculating the composition of the surface layer of ideal multicomponent solutions [308, 309]:

$$x_2^{\omega} = x_2 - \frac{x_2(1-x_2)\omega}{RT}\left(\frac{\partial \sigma}{\partial x_2}\right)_{\frac{x_2}{x_1}} \tag{6.7.1}$$

where x_i^{ω} is the molar fraction of the i-th component in the surface layer; x_i is the molar fraction of the i-th component in the volume; ω is the molar area; σ is surface tension; R is the universal gas constant; T is temperature.

The derivative $\left(\frac{\partial \sigma}{\partial x_2}\right)_{\frac{x_2}{x_1}}$ is taken for the section of the concentration angle.

Usually, equation (6.7.1) is used to determine the composition of the surface layer of multicomponent solutions on the basis of the concentration dependence of the surface tension. However, the equation may also be used to solve an inverse problem, *i.e.*, determination of the surface tension isotherms using the data on the composition of the surface layer.

For the binary systems, the dependence of x_2^ω on x_2 may be described by the linear function [19, 20]:

$$\frac{x_2^\omega}{x_1^\omega} = m\frac{x_2}{x_1} \qquad (6.7.2)$$

Here m is a constant coefficient which is independent of the composition of the solution.

Using the methods described previously, we calculated the composition of the surface layer of experimentally examined ternary systems. Analysis of these results shows that an equality similar to equation (6.7.2) may also be written for ternary systems. However, this is possible only the composition changes in the secant section. Therefore, in deriving surface tension isotherms equations for the ternary systems, we examined this method of changing the composition. In the case of ternary systems, equation (6.7.2) was replaced by the equation

$$\frac{x_i^\omega}{1-x_i^\omega} = m\frac{x_i}{1-x_i} \qquad (6.7.3)$$

x_2^ω is expressed from (6.7.2) or (6.7.3) and substituted into (6.7.1). In this case, regardless of the number of components, we obtain the following differential equation for both the binary and multicomponent systems:

$$d\sigma = -\frac{m-1}{1+(m-1)x_i}\frac{RT}{\omega}dx_i \qquad (6.7.4)$$

In the case of multicomponent systems, an additional condition, which determines the method of variation of the composition, is superimposed on the differential equation (6.7.4). In this case, we examined the variation the composition in the beam sections.

Equation (6.7.4) includes the molar area which depends on the composition of the solution. Therefore, this equation can be integrated only under specific assumptions regarding the dependence of ω on composition. The most suitable variants will be examined.

1. It is assumed that the molar area does not depend on the composition of the solution. In this case, integration of (6.7.4) for ternary systems gives the following equation:

$$\sigma=\sigma_{12}-\frac{RT}{\omega}\ln\left[1+x_3\left(\exp\frac{\sigma_{12}-\sigma_3}{RT}\omega-1\right)\right] \text{ where } \omega=\frac{\omega_1+\omega_2+\omega_3}{3}=\text{const} \quad (6.7.5)$$

where σ_{12} is the surface tension of the two-component solution 1–2.

Equation (6.7.5) for x_1=1 or x_2=1 changes to the well-known equation for the surface tension isotherms of the lateral binary systems 1–3 or 2–3.

2. The concentration dependence of the molar area may also be presented in the form of the additive function $\omega=\sum_{i=1}^{n}\omega_i x_i$. Consequently, integrating (6.7.4) we obtain:

$$\sigma=\sigma_{12}-\frac{RT(m-1)}{A(m-1)+B}\ln\frac{[1+(m-1)x_3]A}{A-Bx_3} \quad (6.7.6)$$

Integration is carried out from the binary alloy 1–2 to the pure component 3.

Taking into account that σ = σ_{12} at x_3=0, from equation (6.7.6) we obtain:

$$\sigma_3=\sigma_{12}-\frac{RT(m-1)}{A(m-1)+B}\ln\frac{mA}{A-B} \quad (6.7.7)$$

where

$$A=\frac{k\omega_1+\omega_2}{k+1}, \quad B=\frac{\omega_2-\omega_3+(\omega_1-\omega_3)k}{k+1}; \quad k=\frac{x_2}{x_1}$$

Parameter m in equation (6.7.7) is calculated by the iteration method.

1. The more area should be the additive function of the composition of the surface layer, if the partial–molar surfaces are equal to the molar surfaces of the pure components. We examined this assumption. It is assumed that

$$\omega=\sum_{i=1}^{n}\omega_i x_i^{\omega}, \text{ i.e. } \omega_i=\varpi_i, \quad (6.7.8)$$

where x^{ω} is the molar fraction of the component in the surface layer, ϖ_i is the partial–molar surface, ω_i is the molar surface of the i-th component.

In chapter 5, the results were presented of calculations of the thermodynamic parameters of the surface layer of the experimentally examined binary and multicomponent systems. The results show that the dependence of the molar area on the composition of the surface layer is considerably closer to the additive straight line than its dependence on the composition of the melt. According to the definition of the molar area, this result is evident in the case in which the partial–molar areas are equal to the molar areas of the pure components.

Equation (6.7.8) is substituted into (6.7.4) followed by integration of this equation. This gives the relatively simple equation for the surface tension isotherms of multicomponent systems:

$$\sigma = \sigma_{ijk} - \frac{RT(m-1)}{m\omega_n - \omega_{ijk}} \ln[1 + \left(\frac{\omega_n}{\omega_{ijk}} m - 1 \right) x_n] \qquad (6.7.9)$$

where σ_{ijk}, ω_{ijk} is the surface tension and molar surface of the i-th component melt to which the n-th component is added. In the case of ternary systems, it is the surface tension and molar surface of the binary alloy 1–2 to which the third component is added. This equation will be written for ternary systems:

$$\sigma = \sigma_{1-2} - \frac{RT(m-1)}{m\omega_3 - \omega_{1-2}} \ln[1 + (\frac{\omega_3}{\omega_{1-2}} m - 1) x_3] \qquad (6.7.10)$$

The resultant equations differ greatly from the previously available equations by the fact that they include the values of the surface tension of not only pure components but also lateral binary systems. Therefore, the resultant equations describe more accurately the concentration dependence of the surface tension of multicomponent solutions. Since the literature contains reliable data on the surface tension of the majority of binary systems, these equations may be used widely for calculations in practice.

4. The concentration dependence of the molar area may be determined using a more accurate thermodynamic equation proposed by Butler:

$$\mu_i^{\omega} - \mu_i = \sigma \bar{\omega} \qquad (6.7.11)$$

where μ_i^{ω} and μ_i are the chemical potentials of the i-th component in the surface layer and in the volume of the solution, respectively; $\bar{\omega}_i$ is the partial–molar area of the i-th component.

Taking into account the well-known identity $\omega = \sum_{i=}^{n} \bar{\omega}_i x_i$, from this equation we obtain the following expression for the molar area:

$$\omega = \frac{1}{\sigma} \sum_{i=1}^{k} \gamma_i x_i^{\omega} + x_i^{\omega} RT \ln \frac{x_i^{\omega}}{x_i} \qquad (6.7.12)$$

For a two-component system, solving together (6.7.4) and (6.7.12), we have:

$$\frac{d\sigma}{\sigma}=-\frac{(m-1)RT}{\gamma_1+(\gamma_2 m-\gamma_1+mRT\ln m)x_2-RT((1+(m-1)x_2)\ln\left[1+(m-1)x_2\right]}dx_2 \tag{6.7.13}$$

where $\gamma_i=\sigma_i\omega_i$ is the molar surface tension.

Integration of equation (6.7.13) faces considerable difficulties. However, the calculations show that $\gamma_1+(\gamma_2 m-\gamma_1+mRT\ln m)x_2 \gg RT(1+(m-1)x_2\ln[1+(m-1)x_2])$. Therefore, the last term in the denominator may be ignored. Consequently, from (6.7.13):

$$\frac{d\sigma}{\sigma}=-\frac{(m-1)RT}{\gamma_1+(\gamma_2 m-\gamma_1+mRT\ln m)x_2}dx_2 \tag{6.7.14}$$

Integrating (6.7.14), after simple transformations we obtain the following equation for the surface tension isotherms of the binary systems:

$$\sigma=\sigma_1\left[\frac{\gamma_1}{\gamma_1+(\gamma_2 m-\gamma_1+mRT\ln m)x_2}\right]^{\frac{(m-1)RT}{\gamma_2 m-\gamma_1+mRT\ln m}} \tag{6.7.15}$$

Parameter m in the above equation is calculated by the iteration method. The obvious conditions are considered, i.e., $\sigma=\sigma_2$ at x_2=1.

The differential equation (6.7.4) is also valid for multicomponent systems. A three-component system will be investigated. We assume that the composition changes in the beam section given by the equation $K=\frac{x_1}{x_3}=\text{const}$. It is assumed that in this method of the variation of composition the condition $K^{\omega}=\frac{x_1^{\omega}}{x_3^{\omega}}=\text{const}$ is fulfilled. Consequently, because of these assumptions, it is possible to integrate equation (6.7.4) if the dependence of the molar area on composition is available. We shall use equation (6.7.12) for these purposes. Substituting (6.7.12) into (6.7.4), after simple transformation we obtain:

$$\frac{d\sigma}{\sigma}=-\frac{(m-1)RTdx_2}{A+(\gamma_2 m-A+mRT\ln m)x_2+((1+(m-1)x_2)RT\ln[1+(m-1)x_2)} \tag{6.7.16}$$

where

$$A=\frac{\gamma_1 k^{\omega}+\gamma_3+k^{\omega}RT\ln\frac{k^{\omega}}{k}}{k^{\omega}+1}+RT\ln\frac{k+1}{k^{\omega}+1}$$

Integration of equation (6.7.6) is associated with considerable difficulties. However, calculations carried out for real systems show that

$$A+(\gamma_2 m-A+mRT\ln m)x_2 >> ((1+(m-1)x_2)RT\ln[1+(m-1)x_2]$$

Consequently, in integration, the term in the denominator may be ignored. Integration of (6.7.16) on the basis of this assumption produces the following expression for the surface tension of ternary systems:

$$\sigma=\sigma_{1-3}\left[\frac{\gamma^*}{\gamma^*+(\gamma_2 m-\gamma^*+mRT\ln m)x_2}\right]^{\frac{(m-1)RT}{\gamma_2 m-\gamma^*+mRT\ln m}} \quad (6.7.17)$$

Where

$$\gamma^*=\frac{\gamma_1 k^\omega+\gamma_3+k^\omega RT\ln\frac{k^\omega}{k}}{k^\omega+1}+RT\ln\frac{k+1}{k^\omega+1}$$

Equation (6.7.17) at x_1=1 and x_3=1 changes to the equation for the isotherms of surface tension of the lateral binary systems 1–2 and 2–3, respectively:

at x_1=1

$$\sigma=\sigma_1\left[\frac{\gamma_1}{\gamma_1+(\gamma_2 m-\gamma_1+mRT\ln m)x_1}\right]^{\frac{(m-1)RT}{\gamma_2 m-\gamma_1+mRT\ln m}} \quad (6.7.18)$$

at x_3=1

$$\sigma=\sigma_3\left[\frac{\gamma_3}{\gamma_3+(\gamma_2 m-\gamma_3+mRT\ln m)x_3}\right]^{\frac{(m-1)RT}{\gamma_2 m-\gamma_1+mRT\ln m}} \quad (6.7.19)$$

Equation (6.7.17) was used for calculating the values of σ for sodium–caesium–potassium ternary systems, in the sections Na:Cs = 6:1, 2:1, 1:1, 1:2, 1:6; indium–tin–lead, indium–tin–gallium, thallium–lead–bismuth, in the sections In:Sn = 9:1, 1:1, 1:9 and Tl: Pb = 9:1, 1:1, 1:9. The calculated values of surface tension, with the exception of the section Tl:Pb = 1:9, are in agreement with the experimental data in the experiment error range. The calculated values for the Pb–Bi the binary system and in the section Tl:Pb = 1:9 differed by 2% in comparison with the experimental values. As an example, the results of calculations for several systems are presented in Fig. 6.31 and Fig. 6.32.

Equation (6.7.15) was used for calculating the surface tension isotherms of 16 binary systems. The results of the calculations show that for a number of binary systems, the calculated values coincide with the experimental data in the experiment error range. The binary systems, for which calculations were carried out, may be divided into three groups.

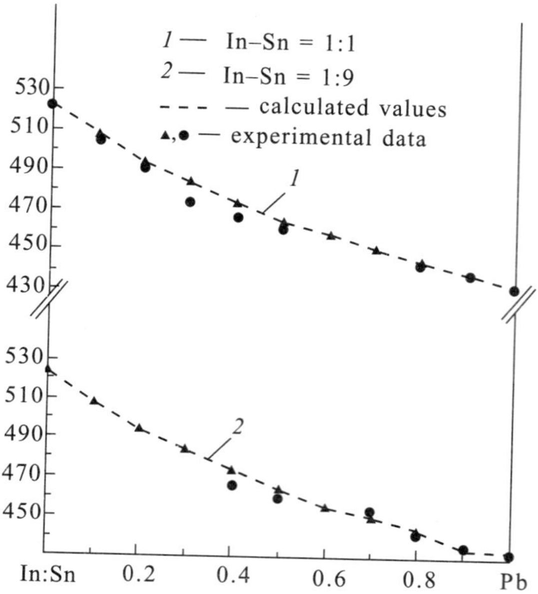

Figure 6.31. Surface tension isotherms of the indium–tin–lead system at a temperature of 723 K.

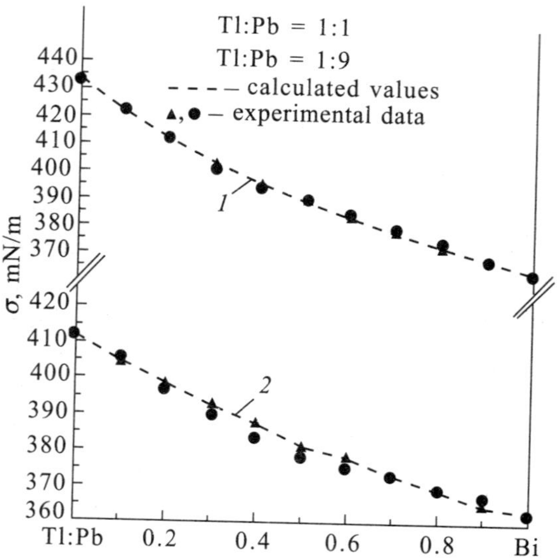

Figure 6.32. Surface tension isotherms of the thallium–lead–bismuth system at a temperature of 723 K.

The first group includes the systems whose surface tension isotherms are described by the proposed equation with the accuracy within the experimental error range. These are sodium–rubidium, sodium–potassium, rubidium–potassium, rubidium–caesium, caesium–potassium, cadmium–lead, indium–gallium, lead–bismuth and tin–thallium binary systems.

The second group consists of the binary systems in which the theoretical isotherms of σ described quantitatively the experimental curves, but in this case there are deviations slightly greater than the error of measurement of surface tension. These are binary systems caesium–sodium, indium–zinc, thallium–bismuth, indium–gallium, tin–gallium.

The third group includes the systems with extrema on the isotherms of σ. The proposed equation does not describe these isotherms even quantitatively. This group includes indium–tin and thallium–lead binary systems.

Some of these results are presented in Fig. 6.33–6.36.

As indicated by the graphs, the proposed equation describes with sufficient accuracy the concentration dependence of the surface tension of binary systems. The advantages of the proposed equation include: a) the simple mathematical form of the equation; b) the equation includes physical quantities which may be determined quite easily by experiments.

Thus, the resultant equations describe satisfactorily the surface tension isotherms of both binary and multicomponent systems. The results of the calculations show that the equation, obtained by taking into account strictly the concentration dependence of the molar surfaces, describes far more efficiently the surface tension isotherms of the binary solutions.

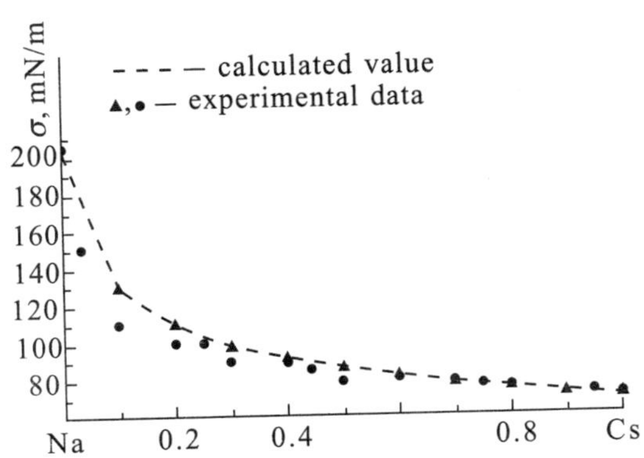

Figure 6.33. Surface tension isotherms of the Cs–Na system at T = 373 K.

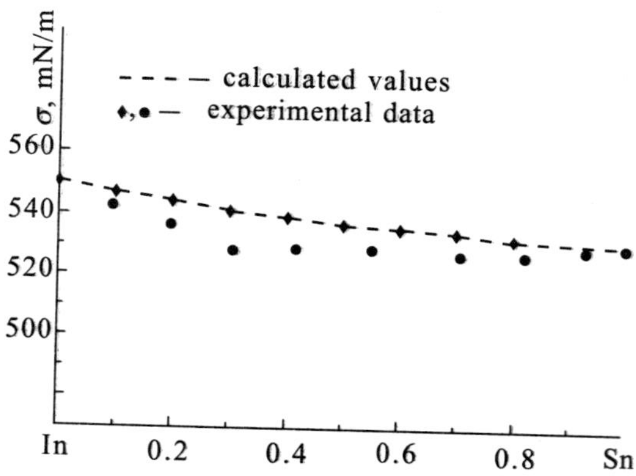

Figure 6.34. Surface tension isotherms of the In–Sn system at $T = 723$ K.

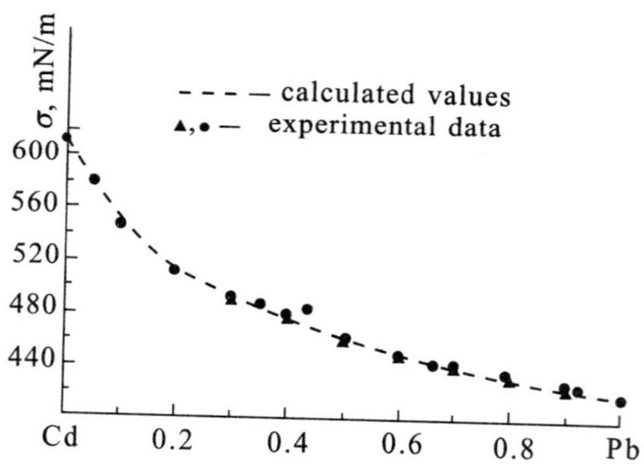

Figure 6.35. Surface tension isotherms of the Cd–Pb system at $T = 723$ K.

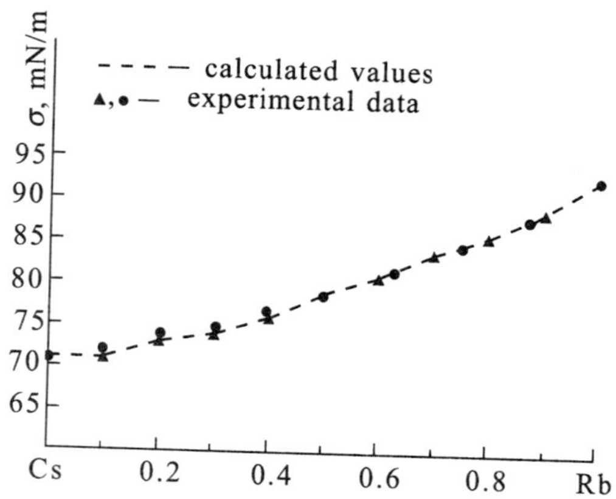

Figure 6.36. Surface tension isotherms of the Cs–Rb system at $T = 373$ K.

Conclusions

1. Using the data on the concentration dependence of the surface tension and molar volumes, the methods described in chapter 1 and 2 were used to calculate the thermodynamic parameters of the surface layer (adsorption, composition and effective thickness of the surface layer, molar and partial–molar surfaces of components) of the experimentally investigated melt.

The adsorption processes in the discussed systems are characterised by various phenomena, such as concentration buffering, suppression of adsorption of some components by others, the independence of the adsorption on the content of the inactive component. In all investigated systems, the dependences of the molar surfaces on the composition of the surface layer are close to the additive dependence, and the phenomenon of the concentration inversion of the surface activity of the component were in the presence of the third component of this component changes from inactive surface-active, was also detected.

2. The comparative analysis of the results of the calculations showed that the isotherms of the parameters of the surface layer in multicomponent and lateral binary systems are closely linked with each other. The isotherms of the surface properties in the beam sections are similar in the form and occupied and intermediate position between their isotherms in the appropriate binary systems.

3. The partial–molar surfaces of the components in the investigated ternary and quaternary systems are equal to their molar surfaces. This is determined by the fact that these solutions are close to ideal. This is also indicated by the isotherms of the molar volumes, close to addictive straight lines.

4. The concentration dependence of the thermodynamic parameters of the surface layer, including the adsorption phenomena in the multicomponent melts, can be predicted on the basis of the isotherms of these properties in the lateral binary systems.

Analysis of the literature data on the surface tension and molar volumes confirms the conclusion according to which the dependences of these quantities on the composition of the multicomponent solution do not contain any special features not predicted by the isotherms of these properties of the appropriate binary systems.

5. A method of predicting the physical–chemical properties of the multicomponent solutions on the basis of the appropriate binary systems has been proposed. The method has been used to calculate the surface tension and molar volumes of 16 ternary and two quaternary systems. Comparison of the calculated values with the experimental data shows good agreement of these results both in the systems with smooth surface tension isotherms and molar volumes and with complicated isothermal surfaces of this properties. Taking this into account, the method has been recommended for evaluating the surface tension of multicomponent melts with any number of the components.

The main results of the investigations

1. The Gibbs adsorption equation was used for deriving systems of linear equations and developing a method of calculating the adsorption of components in relation to different positions of the dividing surface in the multicomponent two-phase fluid system using the dependence of interfacial tension and chemical potential than composition. Comparative analysis of various methods of varying the composition of the solution has been carried out and the methods more suitable for practice have been indicated.

For systems, where one of the phases is an ideal multicomponent solution, we derived explicit equations for the adsorption of components. The equations for different variants of adsorption according to Guggenheim and Adam may be greatly simplified if the content of the components in one of the phases is ignored (the ideal multicomponent solution–saturated vapour interface). The form of the equation for the relative Gibbs adsorption does not change in this case.

The concepts of the surface and limiting surface activity of a component in multicomponent solutions have been described more accurately. For example, the following equation has been proposed for the limiting surface activity of a component in a multicomponent system:

$$G_{0m} = \lim_{x_m \to 0} \left(\frac{\partial \sigma}{\partial x_m} \right)_{K_{ir}}$$

2. The general thermodynamic relationships between the values of adsorption on different dividing surface is of a multicomponent two-phase system have been determined. These relationships may

be used to determine the value of adsorption of the component in relation to any position of the dividing surface, if it's value on one of the surfaces is available. It has been shown that in the diluted multicomponent solutions, as in two-component solutions, the adsorption of dissolved matter depends only slightly on the position of the dividing surface.

The resultant equations were used to develop an adsorption method of determination of the molecular mass of surface-active substances. The method is protected by an Author's certificate.

3. Using the fundamental thermodynamic equations, relationships have been determined for the multicomponent solution-saturated vapour two-phase equilibrium system, and a method has been developed for calculating the main to learn its parameter the surface layer: the molar surface, the composition of the surface layer and its thickness.

The concept of the partial–molar surface of the component has been examined on the basis of the thermodynamic method of the finite thickness layer.

A more general and accurate equation has been derived for the partial–molar surfaces of components. The resultant equation was analysed in comparison with the known approximate relationship. On the basis of theoretical analysis and the calculate the results it has been shown that the isotherms of partial–molar surfaces, obtained using these equations, may qualitatively different from each other. These differences do not have any significant effect on the surface tension isotherms, obtained using the Popel'–Pavlov simulation–thermodynamic equation.

4. A series of improved devices and experimental systems have been developed for the experimental examination of the physical–chemical properties of multicomponent melts. The devices and experimental equipment may be used to produce (without disrupting thermal vacuum conditions) alloys of different compositions and investigate the temperature dependence of their properties. The devices greatly reduce the experimental time and consumption of expensive metals, whilst maintaining the measurement accuracy.

5. The devices have been used to examine in experiments the surface tension and density of low-melting metals: indium, tin, gallium, lead, bismuth, thallium and six binary systems: indium–tin, thallium–lead, bismuth–thallium, bismuth–gallium, indium–thallium, tin–thallium. The experimental data were used in calculating the temperature dependence of the effective thickness of the surface layer of pure metals. Calculations were carried out for the model

of the homogeneous layer of finite thickness assuming that the mean density of the surface layer at the melting point is 95% of the density of the metal. Calculation shows that in all investigated metals, the thickness of the surface layer increases with increasing temperature. The experimental data showed that the surface tension isotherms of the indium–tin and thallium–lead binary systems characterised by a flat minimum in the mean composition range. The depth of the minimum in both systems decreases with increasing temperature. The position of the minimum on the surface tension isotherms of the thallium–lead system does not change and that of the indium–tin melts is displaced from the range of the mean composition in the direction of tin. The special features are explained by a weak interaction of the different atoms in comparison with atoms of the same type. Conditions have been determined which, if fulfilled, may result in the formation of an extreme (minimum, maximum) on the surface tension isotherms. It is shown that these conditions may also be fulfilled in the ideal solutions to which the indium–tin and thallium–lead melt are close.

The isotherms of σ of the In–Tl, Sn–Tl and Ga–Bi binary systems have been used to obtain qualitatively new results. It has been shown that the isotherms of σ of these melts do not contain the extrema detected previously by other authors.

The following parameters were calculated for all the investigated melt: of the distance between different Gibbs dividing surfaces, adsorption, composition and thickness of the surface layer, molar and partial–molar surfaces. It has been shown that the dividing surface $\Gamma_1=0$ moves away from the surface layer with the increase of the content of the second component. The position of the equimolar dividing surface $\left(\sum_{i=1}^{n}\Gamma_{i(n)}=0\right)$ does not change greatly. It has been shown that in the investigated melt, the density of the molar area on the composition of the surface layer is described by the additive straight-line. This indicates that the surface layer of these melts is close to the ideal state.

6. The values of σ and ρ of seven ternary (In–Sn–Pb, In–Sn–Ga, Tl–Pb–Bi, In–Sn–Tl, In–Pb–Tl, Sn–Pb–Tl, In–Bi–Tl) and two quaternary (In–Sn–Pb–Bi, In–Pb–Bi–Ga) metallic systems have been determined by experiments.

The measurement results show that the molar volumes of the In–Sn–Ga and In–Sn–Tl melt coincide with the additive values within the determination error range. The isotherms of the molar volumes in the beam sections of the In–Sn–Pb and Tl–Pb–Bi systems characterised by positive deviations from the additive straight lines. However, in the magnitude they do not exceed the identical deviations in the binary systems.

A general special feature of the isothermal surfaces of the investigated ternary system is the presence of a 'depression' in the range of the composition adjacent to one of the lateral binary systems. The increase of the content of the third component, the depth of the 'depression' decreases and disappears at a specific concentration of the component. As the surface activity of the third component increases, the effect of the component on the surface tension of the ban this system becomes more efficient.

The ternary In–Bi–Tl and In–Pb–Tl systems and the quaternary In–Pb–Bi–Ga system showed the phenomenon of concentration buffering of surface tension. It has been shown that this phenomenon is also typical of ideal multicomponent systems and does not reflect the special features of the structure and composition of the alloys.

The experimental data for the four-component melts may be used to determine the relationships of the combined effect of components on the surface tension. Investigations were carried out into the combined effect of surface-active lead and bismuth (in the indium–tin–lead–bismuth system) and surface-active bismuth and in active indium (in indium–lead–gallium–bismuth melts) on the special features of the surface tension isotherms of binary systems (the minimum on the isotherms of σ of indium–tin and gallium–lead melts). The results show that the efficiency of the effect of the component on the special features of the surface tension isotherms of the binary system is determined by its limiting surface activity, and no effect of the impurities has been detected in the investigated systems. The combined effect of the component is determined by their total concentration in the melt.

Analysis of the results show that the isothermal surfaces of σ and V of the investigated multicomponent melts not contain any special features detected to various degrees on the isotherms of these properties in the lateral binary systems. Special features of the isotherms of the binary systems are manifested on the isothermal surface of σ. With the increase of the content of the third component, these special features reduced.

7. The methods, developed in chapters 1 and 2, and the experimental data, have been applied to calculate the thermodynamic parameters of the surface layer (adsorption, composition and effective thickness of the surface layer, the molar and partial–molar surfaces of components) of multicomponent melts.

The adsorption processes in the investigated systems are characterised by various phenomena, such as the concentration buffering–suppression of adsorption of some component by others, the independence of the adsorption of the surface-active component on the content of the inactive component.

The dependences of the molar surfaces on the composition of the surface layer in all investigated systems are close to additive, and the partial–molar surfaces of the component are equal to the molar surfaces. This indicates that the solutions are close to ideal.

The phenomenon of the concentration inversion of the surface activity of the component was detected, *i.e.*, in the presence of the third component, this component changes from inactive to surface-active.

8. The critical analysis of the experimental data published in the literature confirms the conclusion according to which the concentration dependence of the surface tension and molar volumes in multicomponent metallic melts determined by the isotherms of these properties in lateral binary systems. This shows that the concentration dependences of the thermodynamic parameters of the surface layer in multicomponent notes may be predicted on the basis of the isotherms of these properties in the lateral binary systems.

A method has been developed for predicting the surface properties of multicomponent solutions on the basis of the properties of appropriate binary systems. The surface tension and molar volumes of 16 ternary and 2 quaternary systems have been calculated. Comparison of the calculated values with the experimental data shows that these results are in good agreement both in the systems with the smooth surface tension isotherms and molar volumes, and with complicated isothermal surfaces of these properties. Taking this into account, the proposed prediction method is recommended for the determination of the surface tension of multicomponent melts.

Table 1. Density (kg/m³) of indium–tin melts. Coefficients of the approximation equation $\rho = \rho_0 - \alpha (T-273)$ and the appropriate confidence ranges at a fiducial probability of 0.95

Molar fraction of tin	$\rho_0 \cdot 10^{-3}$ kg/m³	$\alpha \cdot 10$ kg/m³·K	$\Delta\rho_0 \cdot 10^{-3}$ kg/m³	$\Delta\alpha \cdot 10$ kg/m³·K	$\Delta\rho \cdot 10^{-3}$ kg/m³
0.082	7.136	6.8	0.043	1.3	0.026
0.122	7.151	6.8	0.010	0.0	0.010
0.149	7.103	5.9	0.036	0.9	0.019
0.201	7.096	6.1	0.050	1.4	0.023
0.221	7.100	5.9	0.066	1.8	0.029
0.252	7.081	5.9	0.025	0.7	0.017
0.259	7.122	6.7	0.012	0.3	0.005
0.328	7.064	5.5	0.022	0.7	0.020
0.379	7.034	4.7	0.027	0.8	0.016
0.386	7.039	4.5	0.031	0.9	0.025
0.441	7.067	6.0	0.029	0.9	0.025
0.453	7.055	4.9	0.029	0.9	0.018
0.494	7.121	6.8	0.011	0.3	0.004
0.504	7.088	6.4	0.021	0.6	0.009
0.560	7.142	7.1	0.030	0.9	0.019
0.596	7.089	6.1	0.016	0.5	0.014
0.642	7.094	6.2	0.015	0.5	0.012
0.712	7.124	6.9	0.016	0.5	0.009
0.750	7.121	5.7	0.043	0.9	0.040
0.807	7.172	7.1	0.067	2.1	0.022
0.890	7.134	6.8	0.023	0.7	0.010
0.900	7.157	6.6	0.041	0.8	0.036

Table 2. Surface tension (mJ/m²) of indium–tin melts. Coefficients of the approximation equation $\sigma = \sigma_0 - \beta(T-273)$ and the appropriate confidence ranges at a fiducial probability of 0.95

Molar fraction of tin	σ_0 mJ/m²	β mJ/m²·K	$\Delta\sigma_0$ mJ/m²	$\Delta\beta$ mJ/m²·K	$\Delta\sigma$ mJ/m²
0.041	565.7	0.071	6.2	0.017	4.5
0.082	564.1	0.068	1.9	0.005	1.4
0.122	561.2	0.069	2.2	0.006	1.6
0.149	561.0	0.064	1.1	0.003	0.8
0.200	555.0	0.059	2.1	0.005	1.1
0.201	555.0	0.060	0.0	0.000	0.0
0.221	554.0	0.057	1.8	0.005	1.3
0.252	554.1	0.058	1.1	0.003	0.8
0.259	553.0	0.055	1.5	0.004	0.8
0.328	548.3	0.048	2.3	0.006	1.9
0.348	553.0	0.060	0.0	0.000	0.0
0.379	545.6	0.049	1.8	0.006	1.5
0.386	547.4	0.045	1.4	0.004	1.2
0.441	543.5	0.044	3.0	0.009	2.6
0.453	547.2	0.045	1.7	0.005	1.4
0.494	544.6	0.044	2.3	0.007	2.1
0.504	545.7	0.046	1.8	0.005	0.8
0.563	545.9	0.052	2.1	0.006	1.9
0.596	546.0	0.053	3.6	0.011	2.9
0.642	545.6	0.051	3.2	0.010	2.7
0.712	549.9	0.063	1.3	0.004	0.8
0.750	550.0	0.057	1.6	0.004	0.9
0.807	548.9	0.050	2.7	0.008	1.2
0.890	554.0	0.060	0.0	0.000	0.0
0.900	554.0	0.060	0.0	0.000	0.0

Table 3. Results of measurements of the surface tension (mJ/m²) of the indium–tin system

Molar fraction of tin	433 K	473 K	523 K	573 K	623 K	673 K	723 K	773 K
0.041	-	549	548	547	542	537	533	529
0.082	-	550	548	544	540	537	534	530
0.122	-	548	544	540	536	534	530	527
0.1489	-	548	545	542	539	536	532	529
0.20	-	-	540	537	534	532	528	525
0.2012	-	-	540	537	534	531	528	525
0.229	-	-	538	537	534	532	528	525
0.252	-	542	540	537	534	531	528	525
0.259	-	542	539	536	534	532	529	524
0.328	540	536	537	534	533	530	527	523
0.348	-	539	538	535	532	529	526	523
0.379	537	535	534	531	529	526	523	522
0.386	540	538	536	533	530	527	524	521
0.441	535	535	534	531	528	526	523	520
0.453	540	538	535	532	529	526	523	521
0.494	537	538	534	532	530	527	524	519
0.504	539	535	535	533	528	525	521	519
0.563	537	535	533	531	529	525	523	518
0.596	536	535	533	531	528	524	521	517
0.642	537	535	533	531	529	525	521	519
0.712	-	536	534	531	528	525	522	522
0.750	-	-	535	533	530	527	524	521
0.807	-	-	536	534	532	529	526	524
0.890	-	-	539	536	533	530	527	524

Table 4. Surface tension (mJ/m²) and density (kg/m³) of thallium–lead melts

Molar fraction of thallium	623 K		673 K		723 K		773 K	
	$\rho\cdot10^{-3}$	σ	$\rho\cdot10^{-3}$	σ	$\rho\cdot10^{-3}$	σ	$\rho\cdot10^{-3}$	σ
0.10	10.66	440	10.61	438	10.55	435	10.49	432
0.22	10.67	439	10.64	437	10.60	434	10.55	430
0.25	10.69	438	10.64	436	10.59	433	10.55	429
0.31	10.77	437	10.71	435	10.65	431	10.60	429
0.42	10.80	436	10.76	434	10.67	429	10.63	428
0.50	10.88	437	10.81	434	10.74	431	10.67	429
0.57	10.89	438	10.84	435	10.77	432	10.74	429
0.70	10.98	443	10.90	441	10.86	438	10.80	435
0.75	11.07	448	11.00	445	10.95	443	10.88	440
0.90	11.11	456	11.08	454	11.03	451	11.01	448
0.95	11.16	462	11.11	457	11.06	454	11.03	450

Table 5. Surface tension (mJ/m²) of thallium–bismuth melts. Coefficients of the approximation equation $\sigma = \sigma_0 - \beta(T\text{–}273)$ and the appropriate confidence ranges at a fiducial probability of 0.95

Molar fraction of bismuth	σ_0 mJ/m²	β mJ/m²·K	$\Delta\sigma_0$ mJ/m²	$\Delta\beta$ mJ/m²·K	$\Delta\sigma$ mJ/m²	T_0 K
0.042	444.9	0.070	0.6	0.004	0.6	533
0.085	432.9	0.076	1.6	0.009	1.3	523
0.125	425.6	0.080	1.1	0.007	1.0	523
0.170	415.2	0.084	1.3	0.008	1.2	523
0.238	405.1	0.075	1.8	0.011	1.6	523
0.326	395.0	0.066	1.3	0.009	1.0	523
0.443	392.4	0.067	0.5	0.003	0.4	523
0.547	393.1	0.081	1.1	0.007	1.0	523
0.628	390.0	0.058	1.6	0.009	2.0	490
0.695	389.4	0.075	1.1	0.007	1.0	523
0.764	388.0	0.071	0.8	0.005	0.7	523
0.792	388.5	0.087	1.7	0.009	1.7	523
0.860	379.9	0.062	1.4	0.010	1.9	523
0.961	380.0	0.067	1.2	0.008	1.2	533

Table 6. Density (kg/m³) of thallium–bismuth melts. Coefficients of the approximation equation $\rho = \rho_0 - \alpha(T\text{–}273)$ and the appropriate confidence ranges at a fiducial probability of 0.95

Molar fraction of bismuth	$\rho_0 \cdot 10^{-3}$ kg/m³	$\alpha \cdot 10$ kg/m³·K	$\rho_0 \cdot 10^{-3}$ kg/m³	$\Delta\alpha \cdot 10$ kg/m³·K	$\Delta\rho \cdot 10^{-3}$ kg/m³	T_0 K
0.042	11.096	9.87	0.014	1.2	0.017	573
0.085	11.014	13.1	0.024	2.1	0.029	573
0.125	10.905	12.0	0.026	2.3	0.034	573
0.170	10.822	12.2	0.041	3.4	0.049	573
0.240	10.703	12.7	0.054	4.7	0.066	573
0.326	10.593	12.2	0.031	3.5	0.036	573
0.443	10.504	11.5	0.023	1.5	0.021	523
0.547	10.445	11.6	0.032	2.1	0.030	523
0.628	10.372	9.1	0.038	2.2	0.049	490
0.695	10.316	11.6	0.005	0.3	0.005	523
0.764	10.260	10.7	0.026	1.7	0.024	523
0.792	10.300	11.1	0.031	1.8	0.035	473
0.660	10.178	10.6	0.020	1.8	0.026	523
0.961	10.156	11.88	0.010	0.7	0.010	533

Table 7. Surface tension (mJ/m^2) and density (kg/m^3) of indium–thallium melts

Molar fraction of thallium	523		573		623		673		723	
	ρ	σ	ρ	σ	ρ	σ	ρ	σ	ρ	σ
0.00	7.00	570	6.95	565	6.90	560	6.87	557	6.82	550
0.04	-	-	7.13	545	7.05	538	7.02	534	7.00	529
0.08	-	-	7.28	540	7.23	533	7.19	529	7.15	525
0.12	-	-	7.50	524	7.44	518	7.41	515	7.38	511
0.17	-	-	7.65	508	7.65	504	7.58	505	7.57	501
0.24	8.02	496	7.99	499	7.97	496	7.88	494	7.92	492
0.26	8.25	504	8.18	498	8.07	494	8.05	491	8.01	491
0.28	8.24	503	8.20	499	8.08	493	8.04	490	8.01	491
0.32	8.51	500	8.47	498	8.32	488	8.26	487	8.20	485
0.36	8.82	501	8.71	497	8.55	485	8.44	479	8.42	480
0.43	8.99	490	8.95	488	8.86	483	8.75	477	8.65	473
0.51	9.4	489	9.29	483	9.21	479	9.14	474	9.05	469
0.62	9.81	486	9.76	482	9.63	474	9.58	470	9.52	466
0.71	10.10	479	10.04	475	9.99	473	9.91	467	9.86	464
0.73	-	-	10.15	473	10.15	474	10.05	467	9.97	463
0.79	-	-	10.42	474	10.27	473	10.19	465	10.20	463
0.95	-	-	11.05	472	11.01	470	10.88	464	10.78	460
1.00	-	-	11.24	470	11.16	468	11.08	462	11.00	457

Table 8. Surface tension (mJ/m^2) and density (kg/m^3) of tin–thallium melts

Molar fraction of tin	523 K		573 K		623 K		673 K		723 K	
	σ	$\rho \cdot 10^{-3}$	σ	$\rho \cdot 10^{-3}$	σ	$\rho \cdot 10^{-3}$	σ	$\rho \cdot 10^{-3}$	σ	$\rho \cdot 10^{-3}$
0.81	-	7.789	506.6	7.751	505.0	7.713	503.0	7.674	500.0	7.636
0.72	-	8.169	499.1	8.129	498.3	8.089	496.2	8.049	494.0	8.008
0.66	-	8.422	492.8	8.381	492.2	8.339	490.2	8.298	487.7	8.257
0.58	-	8.759	489.8	8.716	487.1	8.674	486.2	8.631	484.0	8.588
0.52	-	9.012	485.4	8.960	483.3	8.924	481.4	8.881	479.6	8.837
0.48	-	9.180	483.4	9.136	480.8	9.022	479.3	9.047	476.7	9.002
0.38	478.7	9.589	476.4	9.543	474.0	9.497	472.4	9.451	469.8	9.404
0.28	473.4	10.006	471.3	9.958	469.3	9.910	466.2	9.863	464.0	9.814
0.21	-	10.305	469.1	10.256	466.6	10.207	461.8	10.158	459.8	10.108
0.16	-	10.520	465.2	10.470	462.5	10.420	460.7	10.370	457.0	10.319
0.06	-	10.993	465.4	10.881	462.4	10.830	459.6	10.778	456.0	10.725

Table 9. Surface tension (mJ/m^2) and density (kg/m^3) of gallium–bismuth melts

Molar fraction of gallium	573 K		623 K		673 K		723 K		773 K	
	$\rho \cdot 10^{-3}$	σ	$\rho \cdot 10^{-3}$	σ	$\rho \cdot 10^{-3}$	σ	$\rho \cdot 10^{-3}$	σ	$\rho \cdot 10^{-3}$	σ
1.000	5.92	703	5.88	700	5.85	697	5.82	694	5.79	691
0.992	5.98	681	5.95	679	5.93	676	5.89	672	5.84	670
0.980	6.07	637	6.05	635	6.03	635	6.00	631	5.96	629
0.966	6.15	579	6.12	579	6.09	580	6.07	579	6.03	578
0.952	6.25	546	6.21	548	6.18	552	6.15	555	6.13	553
0.936	6.36	508	6.31	511	6.29	516	6.26	518	6.24	515
0.924	6.45	496	6.44	501	6.41	508	6.36	510	6.31	508
0.905	6.60	484	6.56	490	5.52	498	6.49	503	6.48	499
0.895	6.65	470	6.63	478	6.60	485	6.56	491	6.53	492
0.876	6.78	468	6.74	474	6.70	480	6.65	487	6.63	486
0.856	6.93	460	6.90	469	6.86	475	6.82	483	6.78	484
0.836	7.03	458	6.99	470	6.95	473	6.92	479	6.91	480
0.818	7.15	446	7.10	456	7.07	464	7.03	470	7.00	469
0.793	7.32	438	7.30	449	7.25	458	7.21	464	7.18	464
0.754	7.53	435	7.50	441	7.45	448	7.40	453	7.38	452
0.713	7.71	422	7.67	429	7.63	435	7.60	441	7.57	439
0.667	7.92	421	7.87	430	7.82	434	7.78	439	7.75	440
0.611	8.20	408	8.16	412	8.12	415	8.09	417	8.04	415
0.547	8.46	398	8.41	398	8.37	400	8.33	402	8.29	400
0.471	8.72	400	8.68	398	8.65	396	8.62	397	8.58	395
0.356	9.10	382	9.05	380	8.99	380	8.25	382	8.90	379
0.300	9.33	386	9.27	381	9.22	379	9.17	379	9.13	376
0.119	9.82	371	9.77	372	9.70	370	9.64	368	9.60	365
0.0	10.13	374	10.08	371	10.02	367	9.96	364	9.91	360

Table 10. Temperature dependence of the surface tension (mJ/m^2) of gallium–bismuth melts

Molar fraction of gallium	573 K	673 K	873 K	973 K	1073 K	1123 K
1.00	694	686	670	662	654	650
0.85	441	447	437	435	426	420
0.80	436	446	429	421	413	408
0.75	427	442	419	410	404	396
0.66	405	410	403	402	391	385
0.50	387	400	382	377	365	363
0.34	380	378	373	361	349	345
0.25	366	362	351	339	336	328
0.00	367	358	340	331	322	317

Table 11. Density (kg/m³) of indium–tin–lead melts. Coefficients of the approximation equation $\rho = \rho_0 - \alpha(T-T_0)$ and the appropriate confidence ranges at a fiducial probability of 0.95

$x_{In} : x_{Sn}$ ratio	Molar fraction of lead	$\rho_0 \cdot 10^{-3}$ kg/m³	$\alpha \cdot 10$ kg/m³·K	$\Delta\rho_0 \cdot 10^{-3}$ kg/m³	$\Delta\alpha \cdot 10$ kg/m³·K	$\Delta\rho \cdot 10^{-3}$ kg/m³	T_0 K
1:9	0.958	10.541	9.4	0.001	0.1	0.001	573
	0.863	10.214	10.6	0.001	0.1	0.001	573
	0.747	9.840	9.0	0.010	1.3	0.021	523
	0.648	9.432	7.0	0.016	1.4	0.023	523
	0.568	9.222	9.0	0.013	0.8	0.013	523
	0.503	8.994	10.2	0.041	1.7	0.040	523
	0.451	8.779	8.4	0.012	0.8	0.012	523
	0.397	8.574	8.4	0.017	0.9	0.019	523
	0.355	8.463	9.1	0.010	0.7	0.016	523
	0.326	8.273	6.4	0.030	2.3	0.025	523
1:3	0.863	10.220	12.1	0.030	2.4	0.027	573
	0.753	9.811	9.4	0.008	0.7	0.011	573
	0.666	9.619	12.5	0.030	1.9	0.029	523
	0.578	9.266	11.1	0.025	1.5	0.024	523
	0.52	9.05	10.1	0.022	1.4	0.028	523
	0.44	8.731	8.6	0.022	1.3	0.028	523
	0.39	8.482	7.9	0.019	1.3	0.027	523
	0.33	8.286	8.0	0.009	0.8	0.14	523
	0.31	8.205	9.0	0.019	1.3	0.023	523
1:1	0.04	7.162	6.2	0.025	1.4	0.027	473
	0.12	7.513	7.1	0.012	0.7	0.017	473
	0.19	7.757	6.7	0.014	1.1	0.030	473
	0.25	8.000	7.4	0.022	1.6	0.032	473
	0.31	8.266	8.0	0.013	0.8	0.016	473
	0.36	8.476	8.5	0.021	1.2	0.013	473
	0.35	8.444	7.9	0.030	1.5	0.031	473
	0.42	8.677	11.2	0.038	1.9	0.039	473
	0.50	8.927	10.2	0.021	1.7	0.034	523
	0.58	9.231	9.2	0.027	1.4	0.013	523
	0.70	9.600	6.9	0.007	0.7	0.008	573
	0.81	10.017	9.7	0.013	1.1	0.018	573
	0.91	10.431	12.7	0.013	1.3	0.016	573
9:1	0.32	8.347	9.6	0.016	0.8	0.016	473
	0.37	8.540	9.6	0.015	1.0	0.025	473
	0.41	8.680	9.6	0.047	1.3	0.036	473
	0.48	8.898	8.8	0.012	0.6	0.011	473
	0.51	9.014	9.4	0.025	1.3	0.022	473
	0.58	9.197	8.4	0.016	0.8	0.012	473
	0.65	9.516	10.4	0.023	1.5	0.023	523
	0.75	9.886	11.7	0.002	0.0	0.001	523
	0.86	10.184	8.1	0.021	1.0	0.007	523
	0.95	10.528	13.3	0.025	1.3	0.021	573

Table 11, continued

$x_{Pb} : x_{Sn}$ ratio	Molar fraction of indium	$\rho_0 \cdot 10^{-3}$ kg/m³	$\alpha \cdot 10$ kg/m³·K	$\Delta\rho_0 \cdot 10^{-3}$ kg/m³	$\Delta\alpha \cdot 10$ kg/m³·K	$\Delta\rho \cdot 10^{-3}$ kg/m³	T_0 K
1	2	3	4	5	6	7	8
1:13	0.07	7.346	7.9	0.021	1.0	0.007	473
	0.09	7.300	7.0	0.011	0.9	0.024	473
	0.14	7.280	7.6	0.011	0.8	0.17	473
	0.23	7.215	6.1	0.011	0.7	0.018	473
	0.34	7.183	6.3	0.012	0.9	0.023	473
	0.44	7.133	6.5	0.007	0.5	0.017	473
	0.47	7.138	6.1	0.014	0.9	0.019	473
	0.51	7.092	5.3	0.017	1.0	0.027	473
	0.65	7.087	6.6	0.004	0.4	0.013	473
	0.71	7.067	6.3	0.021	1.9	0.047	473
	0.78	7.020	5.6	0.022	1.5	0.047	473
	0.89	7.003	6.5	0.021	0.9	0.028	473
	0.95	7.958	5.8	0.030	1.1	0.036	473
1:3	0.95	7.007	5.8	0.007	0.4	0.009	473
	0.86	7.088	5.8	0.025	0.1	0.023	473
	0.78	7.190	5.0	0.042	1.4	0.030	473
	0.71	7.274	5.4	0.020	1.8	0.019	473
$x_{Pb} : x_{In}$ ratio	Molar fraction of tine	$\rho_0 \cdot 10^{-3}$ kg/m³	$\alpha \cdot 10$ kg/m³·K	$\Delta\rho_0 \cdot 10^{-3}$ kg/m³	$\Delta\alpha \cdot 10$ kg/m³·K	$\Delta\rho \cdot 10^{-3}$ kg/m³	T_0 *K*
1:5	0.93	7.069	7.7	0.015	1.2	0.025	523
	0.81	7.112	6.1	0.017	1.2	0.025	523
	0.72	7.118	4.7	0.027	0.1	0.019	523
	0.64	7.201	6.6	0.007	0.6	0.015	523
	0.58	7.243	6.9	0.013	0.9	0.025	523
	0.51	7.282	6.9	0.015	1.3	0.019	523
	0.48	7.308	7.2	0.014	1.1	0.022	523
3:2	0.91	7.224	4.3	0.035	1.4	0.030	523
	0.41	8.416	6.7	0.012	1.0	0.016	523
	0.54	8.181	6.9	0.018	1.6	0.033	523
	0.59	8.000	7.6	0.019	2.1	0.041	523
	0.48	8.294	6.8	0.011	1.2	0.011	523
	Molar fraction of tin	Constant lead contents (3.35 at.%)					
	0.00	7.04	6.6	0.012	1.1	0.027	523
	0.05	7.052	7.8	0.025	1.8	0.021	523
	0.14	7.112	6.9	0.028	2.8	0.055	523
	0.23	7.082	5.9	0.018	1.6	0.039	523
	0.31	7.083	6.0	0.020	2.5	0.0.052	523
	0.37	7.101	5.8	0.001	0.6	0.012	523
	0.42	7.104	6.7	0.010	1.2	0.024	523
	0.44	7.072	4.8	0.010	1.0	0.015	523
	0.49	7.088	4.5	0.020	0.3	0.006	523
	0.56	7.100	5.6	0.005	0.8	0.017	523
	0.60	7.111	5.9	0.015	1.5	0.031	523
	0.67	7.111	6.0	0.007	0.9	0.014	523
	0.77	7.125	6.7	0.011	1.5	0.032	523
	0.86	7.115	5.0	0.010	1.3	0.024	523

Table 12. Surface tension (mJ/m²) of indium–tin–lead melts. Coefficients of the approximation equation $\sigma = \sigma_0 - \beta (T - T_0)$ and the appropriate confidence ranges at a fiducial probability of 0.95

$x_{In} : x_{Sn}$ ratio	Molar fraction of lead	σ_0 mJ/m²	β mJ/m²·K	$\Delta\sigma_0$ mJ/m²	$\Delta\beta$ mJ/m²·K	$\Delta\sigma$ mJ/m²	T_0 K
1:9	0.958	442	0.052	1.2	0.017	1.9	623
	0.863	447	0.058	1.5	0.012	1.5	623
	0.750	455	0.064	0.5	0.006	0.9	573
	0.65	456	0.042	0.7	0.008	1.2	573
	0.57	463	0.046	0.5	0.005	0.8	573
	0.5	467	0.042	0.3	0.009	1.5	573
	0.45	469	0.046	0.5	0.005	0.8	573
	0.4	471	0.040	1.1	0.012	2.5	573
	0.36	471	0.040	1.6	0.011	1.8	573
	0.33	477	0.020	0.0	0.000	0.0	573
	0.00	536	0.060	0.0	0.000	0.0	523
1:3	0.86	448	0.088	0.6	0.007	0.8	623
	0.75	456	0.070	2.3	0.018	2.9	573
	0.67	460	0.078	1.2	0.009	1.5	573
	0.58	463	0.072	1.4	0.011	1.8	573
	0.52	467	0.049	1.6	0.011	2.4	523
	0.44	470	0.036	1.6	0.008	1.8	523
	0.39	476	0.035	1.4	0.009	1.9	523
	0.33	477	0.022	1.5	0.010	2.2	523
	0.31	478	0.026	2.5	0.011	3.0	523
	0.00	536	0.057	0.6	0.004	0.9	523
1:1	0.91	448	0.075	0.7	0.01	1.4	603
	0.81	453	0.056	2.6	0.010	1.6	573
	0.70	457	0.040	1.1	0.010	1.6	573
	0.53	465	0.062	1.6	0.015	2.4	573
	0.50	466	0.030	1.8	0.010	1.8	573
	0.42	474	0.040	2.5	0.018	3.8	523
	0.35	476	0.030	2.3	0.018	3.7	523
	0.36	474	0.015	2.6	0.017	2.6	523
	0.31	482	0.032	1.6	0.009	2.6	473
	0.25	489	0.022	1.9	0.011	2.9	473
	0.19	496	0.021	2.0	0.012	3.2	473
	0.12	509	0.035	1.3	0.007	2.0	473
	0.04	527	0.049	1.8	0.011	2.4	473
	0.00	539	0.066	1.2	0.007	1.9	473
4:1	0.00	573	0.059	1.1	0.005	1.1	473
	0.27	477	0.023	0.6	0.004	0.9	523
	0.31	475	0.026	0.6	0.005	0.8	573
	0.34	474	0.030	0.6	0.005	0.8	573
	0.37	471	0.036	2.0	0.015	2.9	573
	0.44	466	0.030	0.9	0.008	1.2	573
	0.59	457	0.033	1.3	0.008	1.7	573
	0.66	456	0.046	1.5	0.012	2.0	573

Table 12, continued

	0.95	443	0.074	0.5	0.005	0.7	623
	0.86	448	0.043	1.4	0.015	1.8	573
	0.75	456	0.062	1.6	0.012	2.2	573
	0.65	462	0.056	0.7	0.006	0.9	573
	0.57	466	0.040	1.9	0.015	2.4	573
	0.51	472	0.043	1.0	0.018	1.3	573
	0.48	471	0.030	1.5	0.010	1.9	573
9:1	0.41	480	0.06	1.8	0.010	2.6	523
	0.37	482	0.033	0.9	0.006	1.3	523
	0.32	486	0.038	2.4	0.016	2.4	523
$X_{Pb}:X_{Sn}$ ratio	Molar fraction of indium	σ_0 mJ/m^2	β mJ/m$^2\cdot$K	$\Delta\sigma_0$ mJ/m^2	$\Delta\beta$ mJ/m$^2\cdot$K	$\Delta\sigma$ mJ/m^2	T_0 K
	0.00	515	0.035	1.6	0.009	2.4	523
	0.069	518	0.042	0.9	0.013	3.4	473
	0.14	518	0.043	2.1	0.013	3.4	473
	0.23	518	0.034	1.8	0.011	2.9	473
	0.34	524	0.040	1.8	0.011	3.7	473
	0.44	524	0.044	0.7	0.004	1.4	473
	0.47	524	0.039	1.2	0.008	2.6	473
	0.51	525	0.034	1.7	0.009	2.9	473
1:13	0.60	526	0.051	1.2	0.007	2.4	473
	0.65	532	0.060	1.1	0.007	2.3	473
	0.71	534	0.055	0.6	0.004	1.3	473
	0.78	536	0.055	0.9	0.006	1.9	473
	0.89	543	0.065	-0.7	0.002	0.7	473
	0.95	545	0.060	1.4	0.007	2.1	473
	0.95	545	0.055	1.7	0.011	3.6	473
	0.86	532	0.041	0.8	0.006	1.9	473
1:3	0.78	523	0.021	0.7	0.005	1.6	473
	0.71	522	0.029	2.3	0.013	3.3	473
$X_{Pb}:X_{In}$ ratio	Molar fraction of tin	σ_0 mJ/m^2	β mJ/m$^2\cdot$K	$\Delta\sigma_0$ mJ/m^2	$\Delta\beta$ mJ/m$^2\cdot$K	$\Delta\sigma$ mJ/m^2	T_0 K
	0.93	536	0.066	1.1	0.008	1.7	523
	0.81	527	0.053	0.8	0.007	1.4	523
	0.72	520	0.024	1.4	0.008	2.1	473
	0.64	517	0.039	1.5	0.010	2.6	473
	0.58	515	0.045	1.1	0.007	1.6	523
1:5	0.51	514	0.041	0.9	0.005	1.4	473
	0.48	513	0.042	1.3	0.007	1.9	473
	0.91	522	0.020	0.0	0.000	0.0	523
	0.84	512	0.028	1.0	0.005	1.4	473
	0.80	503	0.021	0.9	0.007	1.5	473
	0.71	496	0.026	0.8	0.005	1.3	473
	0.64	489	0.023	1.3	0.007	2.0	473
	0.59	488	0.026	1.3	0.007	1.9	473
3:2	0.58	484	0.018	1.2	0.008	2.1	473
	0.54	486	0.014	0.8	0.008	1.0	473
	0.48	480	0.028	1.0	0.005	1.4	473
	0.41	480	0.030	1.0	0.004	1.3	473

Table 12, continued

Molar fraction of tin	Constant lead content (3.35 at.%)					
	σ_0 mJ/m^2	β mJ/m^2·K	$\Delta\sigma_0$ mJ/m^2	$\Delta\beta$ mJ/m^2·K	$\Delta\sigma$ mJ/m^2	T_0 K
0.00	543	0.071	1.5	0.009	2.4	473
0.05	538	0.054	2.3	0.014	3.7	473
0.14	537	0.051	2.9	0.018	4.8	473
0.23	534	0.044	1.1	0.007	1.8	473
0.31	532	0.052	1.1	0.007	1.8	473
0.37	530	0.040	1.7	0.010	3.4	473
0.42	531	0.050	0.9	0.006	1.5	473
0.44	527	0.036	1.4	0.009	2.4	473
0.49	527	0.039	0.7	0.003	0.9	473
0.56	527	0.039	1.0	0.006	1.6	473
0.60	528	0.050	0.9	0.006	1.5	473
0.67	527	0.049	2.5	0.015	4.0	473
0.77	529	0.052	1.2	0.008	2.1	473
0.86	527	0.031	1.1	0.008	2.0	473

Table 13. Density (kg/m³) of thallium–lead–bismuth melts. Coefficients of the approximation equation $\rho = \rho_0 - \alpha (T - T_0)$ and the appropriate confidence ranges at a fiducial probability of 0.95

Tl:Pb ratio	Molar fraction of bismuth	$\rho_0 \cdot 10^{-3}$ kg/m³	$\alpha \cdot 10$ kg/m³·K	$\Delta\rho_0 \cdot 10^{-3}$ kg/m³	$\Delta\alpha \cdot 10$ kg/m³·K	$\Delta\rho \cdot 10^{-3}$ kg/m³	T_0 K
1:9	0.000	10.726	12.1	0.025	1.9	0.018	573
	0.045	10.615	8.1	0.044	1.6	0.059	623
	0.104	10.583	9.3	0.028	1.1	0.033	623
	0.148	10.602	12.1	0.019	2.6	0.027	573
	0.194	10.559	13.5	0.048	4.1	0.069	573
	0.223	10.568	10.3	0.041	3.1	0.061	523
	0.272	10.512	10.4	0.034	2.6	0.051	523
	0.318	10.470	9.6	0.017	1.3	0.027	523
	0.373	10.440	10.7	0.032	2.6	0.051	523
	0.429	10.435	13.8	0.018	1.4	0.028	523
	0.483	10.333	9.6	0.038	2.5	0.039	523
	0.471	10.401	10.8	0.059	4.1	0.077	523
	0.500	10.438	11.2	0.52	4.1	0.077	523
	0.558	10.310	11.1	0.029	2.3	0.045	523
	0.619	10.271	11.4	0.015	1.1	0.022	523
	0.682	10.197	9.8	0.030	2.2	0.044	523
	0.732	10.198	10.1	0.039	2.1	0.038	473
	0.798	10.132	11.1	0.026	3.3	0.051	523
	0.877	10.078	10.2	0.014	1.2	0.019	523
	0.945	10.024	10.0	0.027	1.9	0.33	523
1:3	0.000	10.687	11.2	0.036	2.5	0.032	623
	0.063	10.637	9.2	0.027	2.3	0.020	573
	0.016	10.625	12.7	0.004	0.9	0.009	573
	0.183	10.537	11.3	0.024	2.2	0.031	573
	0.234	10.494	10.3	0.022	2.7	0.026	573
	0.285	10.472	12.1	0.016	1.7	0.023	573
	0.326	10.424	12.3	0.024	2.3	0.035	573
	0.363	10.361	9.8	0.001	0.9	0.010	573
	0.419	10.333	10.1	0.010	1.1	0.015	573
	0.449	10.354	9.9	0.026	2.4	0.037	523
	0.451	10.373	10.8	0.027	1.4	0.034	523
	0.507	10.331	12.5	0.046	3.0	0.053	523
	0.595	10.245	11.7	0.028	2.7	0.040	523
	0.523	10.344	14.0	0.032	3.1	0.040	523
	0.672	10.166	9.2	0.017	1.4	0.025	523
	0.782	10.126	10.6	0.026	0.8	0.014	523
	0.840	10.102	10.6	0.047	4.5	0.070	523
	0.929	9.990	11.4	0.063	5.4	0.048	573
1:1	0.000	10.689	13.9	0.010	1.3	0.010	623
	0.030	10.808	7.0	0.030	2.2	0.037	573
	0.062	10.810	14.3	0.020	1.5	0.014	573
	0.091	10.756	11.6	0.018	1.5	0.018	573
	0.123	10.753	14.1	0.035	2.9	0.030	573
	0.180	10.664	14.3	0.021	2.2	0.015	573
	0.237	10.564	12.0	0.018	2.3	0.024	573
	0.298	10.498	11.5	0.039	3.8	0.055	573

Table 13, continued

	0.343	10.450	11.1	0.013	1.1	0.017	573
	0.371	10.419	10.9	0.022	2.5	0.034	573
	0.413	10.381	10.8	0.028	2.8	0.038	573
	0.456	10.339	11.2	0.008	0.6	0.009	573
	0.497	10.280	10.5	0.038	3.6	0.052	573
	0.504	10.312	11.8	0.010	1.0	0.019	573
	0.599	10.245	12.9	0.035	3.2	0.047	573
	0.632	10.165	13.1	0.008	0.8	0.011	573
	0.678	10.165	11.8	0.017	1.9	0.028	573
	0.713	10.151	12.1	0.008	0.9	0.013	573
	0.751	10.135	12.8	0.021	2.2	0.034	573
	0.799	10.084	11.6	0.020	1.7	0.033	573
	0.844	10.045	11.0	0.021	2.0	0.029	573
	0.948	10.007	11.3	0.02	2.0	0.029	573
3:1	0.000	11.132	12.1	0.025	2.0	0.010	573
	0.065	11.038	15.1	0.021	2.9	0.025	623
	0.137	10.917	13.3	0.037	5.2	0.046	623
	0.195	10.872	14.1	0.035	3.9	0.054	573
	0.241	10.718	12.3	0.038	3.9	0.054	573
	0.283	10.731	13.0	0.026	1.8	0.032	623
	0.329	10.644	12.8	0.044	3.3	0.062	623
	0.405	10.552	12.0	0.019	1.4	0.025	623
	0.432	10.456	8.9	0.034	2.5	0.047	523
	0.460	10.414	8.6	0.032	2.3	0.041	523
	0.495	10.472	13.5	0.062	4.8	0.051	523
	0.454	10.443	13.0	0.009	0.7	0.014	523
	0.511	10.363	11.6	0.017	1.2	0.024	523
	0.546	10.338	12.0	0.013	1.0	0.020	523
	0.584	10.303	10.7	0.034	2.6	0.051	523
	0.630	10.274	11.8	0.017	1.3	0.026	523
	0.670	10.232	11.6	0.016	1.3	0.024	523
	0.714	10.187	9.1	0.015	1.7	0.033	523
	0.783	10.145	10.7	0.038	1.1	0.033	523
	0.841	10.106	9.8	0.030	1.5	0.028	523
	0.904	10.061	9.1	0.048	2.9	0.039	523
9:1	0.000	11.087	8.6	0.010	2.8	0.012	573
	0.052	10.962	6.1	0.013	1.1	0.010	573
	0.104	10.900	10.7	0.064	2.5	0.051	573
	0.157	10.856	14.3	0.061	2.4	0.078	573
	0.199	10.716	12.0	0.020	1.7	0.026	573
	0.264	10.596	10.4	0.030	2.7	0.039	573
	0.346	10.497	9.2	0.020	1.7	0.025	573
	0.412	10.467	10.6	0.029	2.1	0.042	523
	0.494	10.404	10.1	0.031	2.8	0.045	523

Table 14. Surface tension (mJ/m²) of thallium–lead–bismuth melts. Coefficients of the approximation equation $\sigma = \sigma_0 - \beta\,(T - T_0)$ and the appropriate confidence ranges at a fiducial probability of 0.95

Tl:Pb ratio	Molar fraction of bismuth	σ_0 mJ/m²	β mJ/m²·K	$\Delta\sigma_0$ mJ/m²	$\Delta\beta$ mJ/m²·K	$\Delta\sigma$ mJ/m²	T_0 K
1:9	0.000	440.1	6.7	1.0	1.3	1.3	623
	0.045	433.6	4.8	1.3	1.7	1.9	623
	0.104	427.2	5.0	0.9	1.4	1.0	623
	0.148	423.5	7.0	2.4	1.9	2.7	623
	0.194	423.0	7.4	1.7	1.4	2.1	573
	0.223	423.2	6.8	1.1	0.8	1.6	523
	0.272	418.4	6.9	0.8	0.6	1.2	523
	0.318	413.8	6.7	1.1	0.8	1.8	523
	0.373	410.5	6.6	1.1	0.9	1.9	523
	0.429	408.5	6.8	1.5	1.2	1.5	523
	0.487	405.8	6.4	0.9	0.6	0.9	523
	0.471	408.5	7.0	2.3	1.9	3.3	523
	0.500	404.3	6.5	0.5	0.5	0.9	523
	0.558	401.5	6.5	1.0	0.7	1.6	523
	0.619	398.7	7.1	1.1	0.8	0.6	523
	0.682	394.0	6.1	0.7	0.5	1.1	523
	0.732	391.3	6.1	0.8	0.6	1.2	523
	0.798	388.1	6.1	0.2	0.1	0.2	523
	0.874	382.0	4.6	0.5	0.5	0.8	523
	0.945	379.2	4.9	0.5	0.4	0.7	523
1:3	0.000	437.1	6.4	3.8	0.3	1.7	623
	0.063	426.9	5.8	2.0	1.6	1.6	573
	0.116	424.6	8.3	2.9	2.3	2.3	573
	0.183	416.1	7.6	1.6	1.5	2.2	573
	0.234	412.4	7.0	0.8	0.8	1.2	573
	0.285	407.4	6.5	1.4	1.4	2.0	573
	0.326	405.8	7.4	2.0	1.9	3.0	573
	0.363	403.6	6.3	1.8	1.3	1.7	573
	0.419	400.5	5.2	0.8	0.4	0.6	523
	0.451	402.0	7.8	0.9	0.9	1.5	523
	0.507	396.4	6.0	1.1	1.0	1.6	523
	0.523	395.0	6.4	1.7	1.5	2.4	523
	0.595	390.0	6.4	1.6	1.5	2.4	523
	0.672	387.7	5.0	0.7	0.5	1.0	523
	0.782	385.8	6.3	0.9	0.6	1.3	523
	0.840	385.8	7.1	0.8	0.6	1.2	523
	0.929	379.1	5.1	1.3	0.8	1.2	523
	0.449	400.2	5.5	1.7	1.2	2.6	523
1:1	0.000	440.0	13.9	1.8	2.8	3.4	573
	0.030	434.5	6.9	1.9	1.6	1.9	573
	0.062	424.5	5.4	1.1	0.9	1.0	573
	0.091	420.5	6.8	0.9	0.7	0.8	573
	0.123	418.0	8.0	0.0	0.0	0.0	573
	0.180	411.5	7.8	2.3	1.5	2.9	573
	0.237	404.5	6.8	0.9	0.7	0.8	573
	0.298	403.0	6.8	1.6	1.0	1.6	523
	0.343	400.3	7.0	0.8	0.5	0.8	523

Table 14, continued

	0.371	397.4	6.4	0.9	0.6	0.9	523
1:1	0.413	395.8	6.0	1.3	0.8	1.3	523
	0.456	393.8	6.4	0.9	0.6	0.9	523
	0.497	392.2	5.6	0.9	0.6	0.9	523
	0.504	392.9	7.0	1.3	0.8	1.2	523
	0.590	391.8	8.0	1.3	0.8	1.3	523
	0.632	389.7	7.4	0.8	0.5	0.8	523
	0.678	388.5	7.4	0.8	0.5	0.8	523
	0.713	387.4	7.2	1.7	1.0	1.8	523
	0.751	386.6	7.0	0.8	0.5	0.8	523
	0.844	381.6	6.0	1.6	1.0	1.6	523
	0.899	381.6	6.0	1.6	1.0	1.6	523
	0.948	382.0	6.8	1.6	1.0	1.6	523
	0.000	452.0	10.0	2.8	1.7	1.3	623
	0.065	430.2	6.8	0.8	0.9	1.1	623
	0.137	420.0	6.1	0.3	0.2	0.2	573
	0.195	416.6	8.5	0.8	0.4	0.7	523
	0.241	409.2	8.1	0.6	0.4	0.6	523
	0.283	406.3	8.5	1.0	0.7	1.5	523
	0.329	401.0	8.1	0.3	0.3	0.6	523
	0.405	391.2	6.3	0.7	1.1	2.3	523
	0.432	392.0	6.8	1.6	0.4	1.0	523
	0.460	388.4	6.0	1.0	0.7	1.6	523
3:1	0.495	391.0	7.8	1.4	1.0	2.1	523
	0.454	390.4	7.8	1.4	1.0	2.2	523
	0.511	386.8	6.9	1.1	0.7	1.6	523
	0.546	387.2	8.0	1.3	0.7	1.7	523
	0.584	385.3	7.5	0.6	0.5	1.0	523
	0.630	385.2	7.6	0.8	0.5	1.1	523
	0.670	384.3	8.1	0.7	0.6	1.2	523
	0.714	383.3	7.1	0.8	0.4	1.0	523
	0.783	381.1	6.6	1.1	0.7	1.7	523
	0.841	380.1	6.3	0.9	0.7	1.2	523
	0.904	377.0	5.5	0.7	0.6	0.8	523
	0.000	456.0	7.9	1.9	1.5	1.5	573
	0.052	432.0	5.9	1.6	1.2	1.2	573
	0.104	425.0	9.3	2.7	2.0	2.1	573
9:1	0.157	413.7	8.6	0.9	0.7	1.0	573
	0.199	405.0	7.6	0.9	0.8	1.2	573
	0.264	395.4	6.7	1.0	0.8	1.1	573
	0.346	385.9	6.8	0.6	0.5	0.8	573

Table 15. Density (kg/m³) of indium–tin–gallium melts. Coefficients of the approximation equation $\rho = \rho_0 - \alpha t$ and the appropriate confidence ranges at a fiducial probability of 0.95

In:Sn ratio	Molar fraction of gallium	$\rho_0 \cdot 10^{-3}$ kg/m³	$\alpha \cdot 10$ kg/m³·K	$\Delta\rho_0 \cdot 10^{-3}$ kg/m³	$\Delta\alpha \cdot 10$ kg/m³·K	$\Delta\rho \cdot 10^{-3}$ kg/m³
9:1	0.139	7.063	7.4	0.015	0.4	0.014
	0.206	6.990	7.0	0.043	1.2	0.041
	0.263	6.936	7.2	0.018	0.6	0.020
	0.308	6.895	6.7	0.014	0.4	0.017
	0.359	6.870	7.1	0.048	1.5	0.059
	0.399	6.840	7.1	0.019	0.6	0.027
	0.431	6.750	5.3	0.048	1.5	0.066
	0.447	6.726	5.9	0.028	0.9	0.040
	0.518	6.654	5.1	0.024	0.8	0.034
	0.579	6.681	6.9	0.027	1.0	0.037
	0.615	6.623	6.3	0.020	0.7	0.027
1:1	0.106	7.045	6.5	0.039	1.1	0.036
	0.200	6.984	6.2	0.026	0.8	0.031
	0.267	6.952	7.2	0.028	0.9	0.029
	0.313	6.889	6.0	0.010	0.3	0.011
	0.388	6.849	6.6	0.024	0.7	0.029
	0.421	6.824	6.7	0.097	0.3	0.011
	0.478	6.774	5.9	0.032	0.9	0.037
	0.523	6.732	6.1	0.030	0.9	0.036
	0.562	6.709	5.8	0.034	1.0	0.039
	0.591	6.623	6.1	0.019	0.6	0.023
	0.644	6.602	4.6	0.029	1.0	0.029
	0.652	6.588	6.5	0.022	0.6	0.017
	0.715	6.516	6.1	0.010	0.3	0.011
	0.783	6.485	6.8	0.031	1.0	0.037
	0.847	6.388	6.2	0.020	0.6	0.024
	0.901	6.336	6.7	0.025	0.9	0.026
	0.962	6.268	6.2	0.031	1.0	0.046
1:3	0.065	7.113	6.7	0.067	1.8	0.029
	0.129	7.058	6.8	0.037	1.1	0.023
	0.202	6.997	6.2	0.055	1.8	0.050
	0.242	6.981	6.9	0.019	0.6	0.016
	0.269	6.969	7.0	0.039	1.2	0.033
	0.318	6.907	6.8	0.018	0.6	0.015
	0.379	6.790	4.8	0.036	1.1	0.029
	0.432	6.771	5.4	0.027	0.8	0.017
	0.501	6.698	5.3	0.043	1.4	0.036
	0.566	6.590	4.3	0.022	0.6	0.013
	0.636	6.596	6.2	0.039	1.3	0.048
	0.597	6.999	7.0	0.023	0.7	0.020
	0.703	6.553	5.9	0.028	0.9	0.031
	0.761	6.546	7.2	0.032	1.0	0.039
	0.824	6.461	6.6	0.034	1.1	0.037
1:9	0.147	6.990	5.6	0.058	1.5	0.032
	0.232	6.952	6.1	0.040	1.0	0.027
	0.294	6.908	6.3	0.055	1.4	0.030
	0.347	6.832	5.0	0.033	0.9	0.023
	0.424	6.813	6.6	0.035	1.0	0.032
	0.487	6.737	5.8	0.029	0.9	0.024
	0.571	6.694	6.9	0.047	1.3	0.024
	0.608	6.634	6.5	0.023	0.7	0.022
	0.629	6.572	5.1	0.034	0.8	0.032
	0.673	6.541	5.6	0.027	0.8	0.025
	0.684	6.536	5.9	0.031	1.0	0.038

Table 16. Surface tension (mJ/m^2) of the indium–tin–gallium system

In:Sn ratio	Molar fraction of gallium	323 K	373 K	423 K	473 K	523 K	573 K	623 K	673 K	723 K	773 K
1:1	0.96	631	645	653	661	666	667	667	668	667	666
	0.90	599	606	614	622	626	628	631	632	633	633
	0.85	573	592	599	604	610	610	612	613	614	613
	0.78	585	585	587	591	595	597	598	599	598	597
	0.72	-	-	578	581	584	584	585	585	584	533
	0.65	567	-	572	575	576	578	577	576	576	574
	0.64	-	-	-	576	578	580	580	578	-	576
	0.59	-	-	566	568	569	570	569	569	569	567
	0.56	-	567	570	570	572	573	572	571	569	567
	0.52	-	562	563	565	565	565	564	563	562	559
	0.48	-	562	564	563	562	563	561	560	558	556
	0.42	-	562	563	562	562	561	559	557	555	552
	0.39	-	560	560	560	560	558	557	555	552	551
	0.31	-	556	556	556	555	554	552	550	547	545
	0.27	-	556	556	555	553	551	548	546	542	539
	0.20	-	553	552	551	550	548	545	543	540	536
	0.11	-	-	547	545	544	542	539	535	531	527
1:9	0.15	-	-	-	-	543	541	538	536	534	532
	0.23	-	-	-	549	548	-	544	541	540	537
	0.29	-	-	-	-	551	550	547	545	542	542
	0.35	-	-	-	554	553	552	551	549	547	547
	0.42	-	-	560	559	558	557	555	553	551	550
	0.49	-	-	562	562	562	561	559	559	557	555
	0.57	-	-	567	567	567	566	566	563	562	560
	0.61	-	-	567	570	571	571	570	569	567	566
	0.63	-	-	568	571	572	574	573	573	571	569
	0.67	-	-	572	574	576	577	577	577	577	576
	0.68	-	-	575	578	580	581	580	580	581	581
9:1	0.14	-	-	559	556	554	551	549	545	537	-
	0.21	-	-	560	559	557	553	551	548	546	542
	0.26	-	562	562	561	559	557	554	551	548	-
	0.31	-	-	563	562	561	558	557	554	551	548
	0.36	-	-	561	560	564	563	560	558	555	552
	0.40	-	567	568	567	567	565	564	561	558	555
	0.43	-	566	566	568	569	568	566	566	568	560
	0.48	568	570	568	570	570	570	569	567	566	565
	0.52	568	567	570	572	572	572	572	570	569	568
	0.58	573	578	579	580	579	579	578	577	575	-
	0.61	575	576	578	578	580	580	581	579	578	-
1:3	0.07	-	-	-	-	540	538	536	532	528	525
	0.13	-	-	-	546	544	540	538	536	531	-
	0.20	-	-	548	548	547	543	541	539	536	534
	0.24	-	-	552	550	548	546	544	542	540	538
	0.27	-	-	555	554	551	549	547	544	541	538
	0.32	-	-	555	554	554	554	552	549	546	-
	0.38	-	-	552	556	555	554	554	553	551	-
	0.43	-	-	-	560	560	559	559	556	554	-
	0.50	-	-	561	562	564	563	563	561	560	559
	0.57	-	-	566	564	566	566	569	567	566	-
	0.60	-	-	573	575	574	574	574	572	571	-
	0.64	-	-	568	571	572	573	572	572	570	-
	0.70	-	574	574	579	585	586	587	586	587	-
	0.76	-	584	588	592	594	598	598	597	596	594
	0.82	-	586	598	604	606	605	605	601	600	599
	0.93	618	624	633	640	648	649	651	652	652	650
	0.96	635	-	653	658	663	664	666	-	664	-
	0.89	-	613	618	624	630	632	634	632	-	632

Table 17. Experimental data for the surface tension (mJ/m^2) of bismuth–indium–thallium melts

No.	Molar fractions			523 K		573 K		623 K		673 K		723 K		760 K	
	X_{in}	X_{sh}	X_{Ti}	σ	$\rho\cdot10^{-3}$	σ	$\rho\cdot10^{-3}$	σ	$\rho\cdot10^{-3}$	σ	$\rho\cdot10^{-3}$	σ	$\rho\cdot10^{-3}$	σ	$\rho\cdot10^{-3}$
1	0.25	0.25	0.5	466	9.108	463	9.055	461	8.996	458	8.952	456	8.306	-	-
2	0.25	0.5	0.25	472	8.041	472	8.000	470	7.956	468	7.912	465	7.867	463	7.846
3	0.5	0.25	0.25	487	8.081	484	8.034	483	7.983	479	7.937	476	7.898	475	7.874
4	0.33	0.33	0.34	-	-	473	8.405	470	8.348	467	8.299	465	8.258	463	0.236
	X_{sh}	X_{Ti}	X_{pb}	**Tin–thallium–lead**											
1	0.25	0.25	0.5	-	-	-	-	451	9.896	447	9.839	444	9.797	442	9.771
2	0.25	0.5	0.25	-	-	-	-	449	10.020	443	9.962	442	9.910	439	9.893
3	0.5	0.25	0.25	-	-	457	9.016	454	8.965	451	8.924	448	8.889	448	8.869
4	0.25	0.45	0.30	-	-	470	10.031	465	10.103	463	9.954	466	9.913	-	-
	X_{in}	X_{Ti}	X_{pb}	**Indium–thallium–lead**											
1	0.25	0.5	0.25	-	-	-	-	468	9.939	464	9.874	461	9.839	458	9.805
2	0.5	0.25	0.25	482	9.111	478	9.080	476	9.036	474	8.982	471	8.940	-	-
3	0.75	0.125	0.125	-	-	503	8.064	501	8.011	499	7.971	497	7.936	-	-
4	0.125	0.75	0.125	-	-	-	-	461	10.625	458	10.574	455	10.512	452	10.491
5	0.33	0.33	0.34	-	-	468	9.755	465	9.714	460	9.637	458	9.610	455	9.566

Table 18. Experimental data for the surface tension (mJ/m^2) of bismuth–indium–thallium melts

Alloy composition molar fractions			423 K	473 K	523 K	573 K	623 K	673 K	723 K	773 K
X_{Bi}	X_{In}	X_{Tl}								
1.00	0.00	0.00	-	-	-	377	374	370	366	363
0.75	0.25	0.00	-	-	392	391	388	384	382	378
0.50	0.50	0.00	418	417	416	414	411	408	405	402
0.25	0.75	0.00	-	-	450	448	445	442	439	436
0.00	1.00	0.00	-	559	552	549	545	541	537	531
0.75	0.00	0.25	-	-	388	384	381	373	370	367
0.50	0.00	0.50	-	-	393	389	385	381	377	373
0.25	0.00	0.75	-	-	401	397	394	390	387	384
0.50	0.25	0.25	-	-	394	390	386	381	379	376
0.33	0.33	0.33	-	-	-	399	397	395	392	389
0.25	0.5	0.25	-	419	419	415	413	411	409	405
0.25	0.25	0.5	-	-	406	404	400	397	394	396
0.00	0.75	0.25	-	-	500	498	496	494	492	490
0.00	0.5	0.5	-	-	489	483	479	474	469	465
0.00	0.25	0.75	-	-		474	472	467	463	460
0.00	0.00	1.00	-	-		470	468	462	457	453

Table 19. Polynomial coefficients (5.1) calculated for indium–tin–gallium, indium–thallium–lead, tin–thallium–lead and bismuth–indium–thallium ternary systems at 623 K

Polynomial coefficients	In–Sn–Tl	In–Tl–Pb	Sn–Tl–Pb	Bi–In–Tl
β_1	542	542	534	373
β_2	534	458	458	545
β_3	458	443	443	469
β_{12}	-40	-72	-64	-276
β_{13}	-72	-90	-102	-148
β_{23}	-64	-34	-34	-114
γ_{12}	0.00	-48	-10.667	234
γ_{13}	-48	-58.7	-82.7	166
γ_{23}	-10.667	13.3	13.3	-85
δ_{12}	32.00	-21.3	-21.33	-176
δ_{13}	-21.33	-29.3	-93.3	-176
δ_{23}	21.33	18.7	18.7	-736
β_{1123}	-346.667	21.3	957.333	-570
β_{1223}	1754.667	133.3	-290.667	630
β_{1233}	-570.667	573.3	682.667	-652

Table 20. Surface tension (mJ/m²) and density (kg/m³) of indium–tin–lead–bismuth melts

Composition in molar fractions				473 K		523 K		573 K		623 K		673 K		723 K		753 K	
X_{Bi}	X_{In}	X_{Sn}	X_{Pb}	ρ	σ	ρ	σ	ρ	σ	ρ	σ	ρ	σ	ρ	σ	ρ	σ
0.00	0.33	0.33	0.34	8.877	426	8.850	475	8.816	474	8.785	472	8.749	470	8.722	469	8.693	468
0.06	0.31	0.31	0.32	8.905	459	8.888	458	8.857	457	8.829	455	8.790	453	8.760	452	8.715	449
0.14	0.28	0.28	0.30	8.937	451	8.907	450	8.884	448	8.860	447	8.827	445	8.798	223	8.755	441
0.22	0.25	0.25	0.28	8.991	440	8.952	439	8.933	436	8.903	435	8.863	433	8.848	431	8.815	429
0.29	0.22	0.22	0.27	9.017	432	8.990	430	8.951	428	8.922	427	8.890	425	8.860	423	8.843	421
0.34	0.21	0.21	0.24	9.062	425	9.020	423	8.996	422	8.965	420	8.925	419	8.905	416	8.875	414
0.40	0.19	0.19	0.22	9.075	420	9.055	419	9.020	418	8.990	415	8.965	413	8.920	412	8.910	410
0.43	0.18	0.18	0.21	9.126	417	9.126	417	9.083	415	-	412	8.966	410	8.922	408	8.926	407
0.50	0.16	0.16	0.18	9.288	416	9.224	415	9.146	411	9.066	408	9.048	406	9.003	404	8.980	402
0.54	0.15	0.15	0.16	-	-	9.292	409	9.286	409	9.177	404	9.132	402	9.075	398	9.044	396
0.57	0.14	0.14	0.15	-	-	9.433	407	9.343	402	9.287	399	9.251	394	9.205	394	9.145	391
0.63	0.12	0.12	0.13	-	-	9.494	400	9.46	398	9.400	395	9.335	392	9.283	389	9.246	387
0.72	0.09	0.09	0.100	-	-	-	-	9.560	393	9.500	388	9.437	385	9.408	383	9.374	381
0.82	0.056	0.056	0.064	-	-	-	-	9.705	386	9.663	384	9.609	382	9.557	378	9.513	376
0.86	0.044	0.044	0.050	-	-	-	-	9.776	383	9.737	381	9.686	379	9.620	376	9.580	373
0.947	0.017	0.017	0.198	-	-	-	-	9.888	379	9.811	376	9.784	374	9.730	371	9.688	368
1.00	0	0	0	-	-	-	-	9.965	376	9.952	375	9.893	371	9.801	367	9.758	365

References

1. Reed R., *et al.*, Properties of gases and liquids [Russian translation]. Khimiya, Leningrad, 1982.
2. Naidich Yu ., *et al.*, Surface properties of fluxes and solids and their applications in materials science, Naukova Dumka, Kiev, 1991.
3. Blinkin V. and Novikov Z., Liquid salt atomic reactors, Metallurgiya, Moscow, 1978.
4. Pantsevich I., *et al.*, Surface phenomena in welding of metals, Naukova dumka, Kiev, 1991.
5. Izmailova V., *et al.*, Surface phenomena in protein systems, Khimiya, Moscow, 1988.
6. Semenchenko V., Surface phenomena in metals and alloys, Gostekhizdat, Moscow, 1957.
7. Physical chemistry of inorganic materials, in 3 volumes, edited by Yeremenko V., Naukova dumka, Kiev, 1988, vol. 2.
8. Rusanov A., *et al.*, Surface separation of substances, Khimiya, Leningrad, 1981.
9. March N. and Tosi M., Movement of liquid atoms, Metallurgiya, Moscow, 1980.
10. Rowlinson J. and Widom B., Molecular theory of capillarity [Russian translation], Mir, Moscow, 1986.
11. Crockston K., Physics of liquid state [Russian translation], Mir, Moscow, 1978.
12. Nizhenko V. and Floka L., Surface tension of liquid metals and alloys, Metallurgiya, Moscow, 1981.
13. Popel S., Theory of metallurgical processes, VINITI, Moscow, 1976.
14. Shebzukhov A. and Khokonov Kh., in: Physics of interface phenomena, KBGU, Nal'chik, 1980.
15. Gibbs J., Thermodynamics. Statistical mechanics [Russian translation], Nauka, Moscow, 1982.
16. Van–der–Waals I.D. and Konstamm F., Course of thermodynamics [Russian translation], vol. I, ONTI, 1936.
17. Van der Waals I.D., Thermodynamische Theorie der Capillaritat, Verh. Kon. Acad., Amsterdam, 1893.
18. Guggenheim A., Thermodynamics, North Holland Publishing Co., Amsterdam, 1967.
19. Rusanov A., Phase equilibria and surface phenomena, Khimiya, Leningrad, 1967.
20. Rusanov A., Phasengleichgewichte und Oberflachenerscheinungen, Academie–Verlag, Berlin, 1978.
21. Chattaraj D., *Indian J. Chem.*, 1981, **20**, No 80, 941–947.
22. Adamson N.A., Physical chemistry of surfaces [Russian translation], Mir, Moscow, 1979.
23. Lopatkin A.A., *Zh. Fiz. Khimii*, **63**, No. 9, 2433–2441.
24. Popel S.I., in: Surface phenomena inmelts and solid phases formed from them, Istiintsa, Kishinev, 1974, 47–75.
25. Semenchenko V.K., About some thermodynamic issues of surface phenomena, in: Wettability and surface properties of fluxes and solids, Naukova dumka, Kiev, 1972.

26. Rusanov A.I., 100th anniversary of Gibbs's theory of capillarity, Modern theory of capillarity, Khimiya, Leningrad, 1980.
27. Abramzon A.A., Surface substances, Khimiya, Leningrad, 1981.
28. Rusanov A.I., *DAN sSSR*, **261**, No.3, 700–733.
29. Hammish L. and Shebzukhov A.A., IN: 7th All–Union Conference, 1, Part 2, Chelyabinsk, 1990, 165–267.
30. Markin V.S. and Kozlov M.M., Langmuir,1989. **5**, No.4., 1130–1132.
31. Arsamikov U.V., *et al.*, in : Adhesion of melts and brazing of materials, 1989, **23**, 10–13.
32. Guggenheim I.A., Modern thermodynamics in accordance with the Gibbs method, GNTI Khim. Lit., Leningrad, 1941.
33. Dadashev R.Kh., *et al.*, *Zh. Fiz. Khim.*, 1982, **56**, No.10, 2492–2495.
34. Dadashev R.Kh. and Yushaev S., *Zh. Fiz. Khim.*, **59**, No.6, 1562–1565.
35. Levichev S.A., *et al.*, Physical Chemistry of contact phase interface, Naukova Dumka, Kiev, 1976.
36. Dadashev R.Kh., Adhesion of melts and brazing of materials, Naukova dumka, Kiev, 1987.
37. Shcherbakov L.M. and Samsonov V.M., Thermodynamics of surface phenomena, KGU, Kalinin, 1986.
38. Vaisburg S.E. and Sheikin A.B., Surface phenomena in melts, Naukova dumka, Kiev, 1968.
39. Hillert M., *Calphad*, 1988, **12**, No.3, 257–259.
40. Butler I.A.V., *Proceedings of the Royal Society*, 1932, **135**, No. 827, 348–375.
41. Lopatkin A.A., Theoretical principles of physical adsorption, Moscow State University, Moscow, 1983.
42. Good R.I., *Pure and Appl. Chem.*, 1976, **48**, No.4, 427–433.
43. Schaj B., *Pure and Appl. Chem.*, 1976, **48**, No.4, 393–400.
44. Schaj B., *Colloid and Polym Sci.*, **260**, No. 119, 888–891.
45. Everett G., *Pure and Appl. Chem.*, 1981, **53**, No.11, 2181–2197.
46. Flad E., in: Gas–solid interface [Russian translation], Mir, Moscow.
47. Dadashev R.Kh., in: Analytical chemists of Northern Causasus to industry, ChIGU, Groznyi, 1989.
48. Sahoo P. and Small W., *Scr. Met. et Mater.*, 1990, **24**, No. 4, 645–649.
49. Nizhenko V.N. and Khilya G.P., in: Structure and properties of metallic and slag melts, 7th All–Union Conference, Part I, Part 2, Chelyabinsk, 1990.
50. Shur A.M., High–molecular compounds, Vysshaya shkola, Moscow, 1981.
51. Kiriyanenko A.A., *et al.*, Method of determining molecular mass of polymers. Certificate of authorship No. 1738375/26–25I3I3, 1972.
52. Guggenheim E. and Pru J., Physico–chemical calculations, Izdatelstvo inostrannoi literatury, Moscow, 1958.
53. Dadashev R.Kh., *et al.*, Method of determining molecular mass of surface substances, Certificate of authorship No. 1460677 of 22 October 1988.
54. Kogan V.B., Heterogeneous equilibria, Khimiya, Leningrad, 1968.
55. Lovett R., *et al.*, *J. Chem. Phys.*, 1973, **58**, No.5, 1880–1885.
56. Bakker G., Kapillaritat und Oberflachenspannung. Handb. d. Experimentalphysik, 6, Leipzig, 1928.
57. Rusanov A.I. and Levichev S.A., in: Surface phenomena in melts, Naukova dumka, 1968.
58. Ibragimov Kh.I. and Savvin V.S., in: Physico–chemical investigations of metallurgical processes, Izd. UPI, Sverdlovsk, 1979.
59. Nal'giev A.G., Investigation of surface tension, density and work function of the

electron in some binary metallic systems, Thesis, Nal'chik, 1975.
60. Aziev S.L. Examination of surface tension and density in metallic melts based on mercury, Thesis, Nal'chik, KGBU, 1975.
61. Khokonov Kh.B., *et al.*, *Dokl. AN SSSR*, 1973, **210**, No.4, 899–902.
62. Rusanov A.I., *et al.*, *Zh. Fiz. Khim.*, 1969, **43**, No.10, 2636–2638.
63. Ibragimov Kh.I. and SavvinV.S., Surface properties of melts, Naukova dumka, Kiev, 1982, 22–24.
64. Ono S. and Kondo S, Molecular theory of surface tension in liquids, Izd. Inostr. Lit., Moscow, 1963.
65. Vaitsman E.V., *Zh. Fiz. Khim.*, 1985, **59,** No.8, 2101–2105.
66. Rusanov A.I., *et al.*, Surface phenomena in liquids and liquid solutions, Izd–vo LGU, Moscow, 1972.
67. Vaitsman E.V., *Zh. Fiz. Khim.*, 1985, **51**, No.2, 803–811.
68. Gruverman S.L., *et al.*, *Metally*, 1986, No.2, 67–71.
69. Pshenitsyn V.I. and Rusan A.I., Surface phenomena in liquids and liquid solutions, Izd. LGU, Leningrad, 1972, 150–157.
70. Huang I.S. and Webb W.W., *J. Chem. Phys.*, 1965, **50**, No. 9, 3677–3689.
71. Lekner I. and Henderson I.K., *Physica*, 1978, **94**, No.3/4, 548.
72. Heyes D.M. and Clarke I.H.R., *J. Chemical Soc. Faraday Transactions,* 11, 1979, **75,** No.9, 1240–1255.
73. Langevin D., *Prog. Colloid and Polym. Sci.*, 1990, **83**, 10–15.
74. Analysis of surface by Auger methods and X-ray electronic spectroscopy [Russian translation], ed. Briggs D. and Stikh M.P., Mir, Moscow, 1987.
75. Zenguil E. Physics of surfaces [Russian translation], Mir, Moscow, 1990.
76. Woodraf D and Delchar T., Modern methods of surface examination [Russian translation], Mir, Moscow, 1989.
77. Stanley G., Phase changes and critical phenomena, Mir, Moscow, 1973.
78. Fisher M., Nature of critical state, Mir, Moscow, 1968.
79. Yovruch I., *Colloids and Surfaces,* 1989, **40**, No.1–2, 85–91.
80. Zhukhovitskiy A.A., *Zh. Fiz. Khim.*, 1944, **18**, No.5–6, 214–233.
81. Zadumkin S.N., *et al.*, Teoreticheskaya i eksperimentalnaya khimiya, 1971, No.1, 34–39.
82. Zadumkin S.N. and Khokonov Kh.B., Physical chemistry of melt surfaces, Metsniereba, Tbilisi, 1977.
83. Hoard T.P. and Melford D.A., *Trans. Faraday Soc.*, 1957, **53**, 315–325.
84. Khilya R.P., in: Surface properties of melts, Naukova dumka, Kiev, 1982.
85. Ofitserov A.A., *et al.*, *Izv. VUZ, Chernaya metallurgiya*, 1966, No.9, 5–9.
86. Prigozhin I. and Defey R., Chemical thermodynamics, Novosibirsk, 1966.
87. Morachevskiy A.G., Thermodynamics of molten metallic and salt systems, Metallurgiya, Moscow, 1987.
88. Petrenko V.E. and Kessler Yu.M., *Izv. VUZ, Khim. Khim. Tekhnol.*, 1985, **28**, No.5.
89. Wagner K., Thermodynamics of alloys, transl. from English, edited by Zhukhovitskiy A.A., Metallurgizdat, Moscow, 1957.
90. Sorin M.V. and Dymov V.E., *Zh. Fiz. Khim.*, 1987, **51**, No.10, 2540–2645.
91. Nemchenko V.P. and Popel' S.I., *Zh. Fiz. Khim.*, 1969, **43**, No.7, 1822–1827.
92. Popel' S.I. and Pavlov V.V., Surface phenomena in melts and solid phases formed from them, Kabard.–Balk. Knizh. Izd., Nal'chik, 1965, 46–60.
93. Popel' S.I. and Zakharava T.V., in: Physico-chemical investigations of metallurgical processes, Izd.UPI, Sverdlovsk, 1989.
94. Shergin L.M., *et al.*, *Tr. Inst. Metallurgii UNTs AN SSSR,* 1971, No.25, 52–61.
95. Shergin L.M., *et al.*, *Zh. Fiz. Khim.*, 1970, **44**, No.1, 360–361.

96. Arsemikov U.V., *et al.*, Adhesion of melts and brazing of materials, Naukova, dumka, 1991, No.25, 26–29.
97. Kononenko V.N., *et al.*, Zh. Fiz. Khim., 1972, **46**, No.6, 1589–1590.
98. Ivashechko Yu.M., *et al.*, Principles of precise measurement of surface energy of melts using the sessile droplet method, Kiev, Naukova Dumka, 1972.
99. Hansen F.K. and Rodsrud G., *Colloid and Interface. Sci.*, 1991, **141**, No.1, 1–9.
100. Ibragimov Kh.I. and Savvin V.S., Methods of investigation and properties of rhe interface of contacting phases, Naukova dumka, Kiev, 1977.
101. Mezdrogina M.M. and Sidorova T.A., in: Physical chemistry of melt surfaces, Metsniereba, 1974, 279–283.
102. Khokonov Kh.B., Surface phenomena in melts and solid phases formed from them, Istnintsa, Kishinev, 1974, 190–201.
103. Kisil I.S., *Koll. Zh.*, **53**, No.4, 642–645.
104. Vertman A.A. and Samarin A.A., Methods of investigation of the propertie of metallic melts, Moscow, 1969.
105. Kunin L.A., Surface phnomena in metals, Metallurgizdat, Moscow, 1955.
106. Khantadze D.V., *et al.*, Some applications of capillarity theory for physical and chemical investications of melts, Mrtsniereba, Tbilisi, 1971.
107. Missol V., Surface energy of interfaces in metals, Metallurgiya, Moscow. 1978.
108. Arsent'ev P.P., *et al.*, Physical and chemical methods of investigation of metallurgical processes, Moscow, 1988.
109. Bliznyukov S.A., *et al.*, *Izv. VUZ, Chernaya Metallurgiya*, 1964, No.7, 227–238.
110. Sokolov V.I. and Popel S.I., in: Surface phenomena in melts and solid phases formed from them, KBGU, Nal'chik, 1965, 216–222.
111. Nizhenko V.I., Surface phenomena in melts and solid phases formed from them, Istnintsa, Kishinev, 1974, 261–267
112. Tavadze F.N., *et al.*, *Zh. Fiz. Khim.*, 1970, **44**, No.11, 2910–2911.
113. Popel S.I., *et al.*, Graphs for calculation of surface tension on the basis of the shae of the sessile droplet, UPI, Sverdlovsk, 1961.
114. Eremenko V.N., *et al.*, in: Experimental equipment and methods of investigation at high temperatures, Moscow, 1959, 285–294.
115. Iodkovskiy S.A., *et al.*, Scientific and technical information on work of TsNIITMash, Moscow, 1965, No.51, 109–118.
116. Kivilis S.S., Methods of measuring density, Standartgiz, Moscow, 1959.
117. Kivilis S.S., Densitometers, Energiya, Moscow, 1980.
118. Tavadze F.N., Inventions. Industrial designs, Certificate of authorship No. 259456 (USSR), 1970, No.2.
119. Nal'giev A.R.–M., *et al.*, Investigation of surface tension, density and work function of the electron of the tin–mercury system, ChIGU, Groznyi, deposited at the VINITI on 10th April 1975 under No.991–75.
120. Pugachevich P.P., in: Surface phenomena in metallurgical processes, Moscow, 1963, 177–192.
121. Pugachevich P.P., in: Surface phenomena in melts in powder metallurgy processes, Naukova dumka, 1963, 422–432.
122. Khokonov Kh.B., *et al.*, *Dokl. AN SSSR*, 1973, **210**, No.4, 899–901.
123. Zamyatin V.M., *et al.*, *Zavod. Lab.*, 1985, No.6, 66–67.
124. Zamyatin Z.M., *et al.*, Certificate of authorship No. 960585 (USSR), Bolshaya Entsiklopediya, 1982, No.35.
125. G.Lang, *et al.*, *Z. Metall.*, 1977, **68**, No.2, 113–116.
126. Carla M., *et al.*, *Rev. Sci. Instr.*, 1991, **62**, No.4, 1088–1092.
127. Pugachevich P.P. and Lazarev V.B., *Dokl. AN SSSR*, 1957, **117**, No. 3, 445–447.

128. Shebzukhov A.A. and Ashkhotov O.G., *Dokl. AN SSSR*, 1984, **274**, No.6, 1427–1430.
129. Zadumkin S.N., *et al.*, Izv. VUZ, *Tsvetnaya Metallurgiya*, 1979, No. 1, 82–85.
130. Ibragimov Kh.I., *et al.*, Surface phenomena in melts and solid phases formed from them, KBGU, Nalchik, 1965.
131. Filippov S.I., *et al.*, Physical and chemical methods of investigation of metallurgical processes, Metallurgiya, Moscow, 1968.
132. Tkhorzhevskiy V.P., Automated analysis of gases and liquids at chemical enterprises, Khimiya, Moscow, 1976.
133. Gauzer S.I., *et al.*, Measurement of mass, volume and density, Izd. Standartov, Moscow, 1972.
134. Basin A.S., *et al.*, in: Gamma method in metallurgical experiments, ITF SO AN SSSR, Novosibirsk, 1981.
135. Betin Yu.P., *et al.*, *Izotopy v SSSR*, 1976, No.8, 11–13.
136. Gart G., Radioisotope measurenment of density of liquids in binary systems, Atomizdat, Moscow, 1975.
137. Dadashev R.Kh., Surface tension and adsorption in multicomponent systems, UPI, Ekaterinburg, 1993.
138. Brazhnikov N.I., *Pribory i Sistemy Upravleniya*, 1976, No. 10, 17–21.
139. Zhukov Yu.P. and Begunov V.N., *Pribory i Sistemy Upravleniya*, 1976, No.9, 28–30.
140. Ibragimov Kh.I., *et al.*, in: Scientific conference on Physics of interfacial phenomena, Nal'chik, KBGU, 1971.
141. Khokonov Kh.B., *et al.*, in: Investigations of surface phenomena in metallic melts, Ordzhinikidze, SOGU, 31–37.
142. Savvin V.S., Surface phenomena of amalgams of thallium, lead and bismuth, Thesis, Groznyi, ChIGU, 1987.
143. Nizhenko V.I., In: Methods of investigation and properties of the interface of contacting phases, Kiev, Naukova dumka, 1977.
144. Vukalovich M.P., *et al.*, Thermophysical properties of mercury, Moscow, Standard Publishing House, 1971.
145. Nal'giev A.G., Ibragimov Kh.I., *Zh. Fiz. Khim.*, 1974, **48**, 1289.
146. Ibragimov Kh.I., Surface phenomena in melts based on mercury and metals of groups III–IV, Thesis, Kiev, IPM, 1980.
147. Alchigirov B.B., Surface tension, density and work function of the electron of some binary metallic melts, Thesis, Nal'chik, KBGU, 1974.
148. Grawley A.P., *Int. Met. Revs.*, 1974, **19**, 32–48.
149. Knobler C.M., Pings C.I., *J. Chem. Eng. Data*, 1965, **10**, No.2, 129–130.
150. Ibragimov Kh.I., Savvin V.S., Dadashev R.Kh., *Zh. Fiz. Khim.*, 1966, **40**, No.4, 957–959.
151. Ibragimov Kh.I., Pokrovskii N.L., Pugachevich P.P., *Zh. Fiz. Khim.*, 1966, **40**, No. 4, 957–959.
152. Ibragimov Kh.I., Dadashev R.Kh., in: Investigation of surface phenomena in metallic melts, Ordzhonikidze, SOGU, 1975, 62–68.
153. Alchigirov B.B., Surface tension, density and work function of the electron of some binary metallic melts, Thesis, Nal'chik, KBGU, 1974.
154. Ibragimov Kh.I., Dadashev R.Kh., in: Investigation of surface phenomena in metallic melts, Ordzhonikidze, SOGU, 1989, 68–71.
155. Aziev S.L., Examination of surface tension and density in metallic melts based on mercury, Thesis, Groznyi, ChIGU, 1975.
156. Sen'kin I.F. and Spiridonov Yu.S., *Elektronnaya phomyshlennost*, 1974, No.5, 76–

78.
157. Norton P.I., Gas permeation through the vacuum envelope, Trans. 8th National Vacuum Symposium, 1961, New York, Pergamon Press, 1962.
158. Toybert R., Evaluation of the accuracy of measurement results, Moscow, Energoatomizdat, 1988.
159. Hudson D., Statistics for physicists [Russian translation], Moscow, Mir, 1970,
160. Şavvin V.S., Surface phenomena of amalgams of thallium, lead and bismuth, Thesis, Groznyi, ChIGU, 1978.
161. Kalosha Z.K., *et al.*, Mathematical processing of experimental results.
162. Rebinder P.A. and Vladovets V.N., Surface tension, Physical Encyclopeadical Dictionary, vol. 4, Sov. Entsikl., Moscow, 1965, 56–59.
163. Yatsenko S.P., et al., Properties of gallium in water solutions and alloys, Tr. Inst. Khimii UF AN SSSR, Sverdlovsk, 1962, No.12.
164. Borishchanskii V.M., Liquid metal heat carriers, Atomizdat, Moscow, 1967.
165. Alchagirov B.B. and Osiko T.P., in: Physics of surface phenomena, Nal'chik, KBGU, 1985, 79–90.
166. Ibragimov Kh.I., Savvin V.S., in: Physical–chemical investigations of metallurgical processes, Sverdlovsk, UPI, 1980, 61–66.
167. Khilya G.P., *et al.*, in: Adhesion of melts and brazing of materials, 1987, No. 19, 20–23.
168. Asramiskov U.V., et al., Adsorption phenomena in melts of the indium–gallium system, Cherkassy, 1989, deposted at ONIITE, No. 100-XII, 89.
169. Kornilov A.N., *Zh. Fiz. Khim.*, 1967, **41**, No.12, 3096–3102.
170. Kornilov A.N., *Zh. Fiz. Khim.*, 1967, **41**, No.12, 3102–3106.
171. Tokmanova A., *et al.*, Izv. AN SSSR, Neorg. Mater., 1981, **17**, No.11, 2095–2099.
172. Yeazey S.D., Roe W.C., *J. Mater. Sci.*, 1972, **7**, No.4, 445–446.
173. Karashaev A.A., *et al.*, *Zh. Fiz. Khim.*, 1968, **41**, No.3, 654–658.
174. Nizhenko Z.I., *et al.*, *Ukr. Khim. Zh.*, 1965, No.6, 559–563.
175. Timofeevicheva O.A. and Pugachevich P.P., *DAN SSSR*, 1960, **134**, No.4, 840–843.
176. Chabra Rojendra P., *High Temp - High Pressures*, 1990, **22**, 171–173.
177. Abbaschian G.I., *Less. Common Metals*, 1975, **40**, No.3, 329–333.
178. Smitlz K.G., Metals [Russian translation], Moscow, Metallurgiya, 1980.
179. Zinov'ev Z.E., Thermophysical properties of metals at high temperatures, a handbook, Moscow, Metallurgiya, 1989.
180. Passerone A., Ricei E., Sangiargi R., *J. Mater Sci.*, 1990, **25**, No.10, 4266–4272.
181. Stankus S.V. and Tyagel'skii P.V., *Rasplavy*, 1991, No.2, 14–19.
182. Dritts M.E. (ed), Properties of elements, a handbook, Moscow, Metallurgiya, 1985.
183. Table's of physical quantities, Moscow, Atomizdat, 1976.
184. Defau R. and Prigogine I., Surface tension and adsorption, London, Longmans Green and Co Ltd, 1966.
185. Johnson M.D., Hutchinson P., March M.N., *Proc. of the Royal Society*, 1964, **282**, No.1389, 283–302.
186. Ambrok A.G., and Kalashnikov E.V., *Rost of krystalov*, 1990, **18**, 5–17.
187. Ambrok A.G., *et al.*, *Poverkhnost'*, 1989, No.4, 137–142.
188. Maiboroda V.P., *Rasplavy*, 1991, No.3, 117–121.
189. Koval'chuk V.F., The density and surface properties of liquid alloys indium–tin, indium–bismuth, thallium–lead, and thallium–tin, Sverdlovsk, UPI, 1969.
190. Kononenko V.N., *et al.*, *Zh. Fiz. Khim.*, 1972, **46**, 1589–1590.
191. Koval'chuk V.F. and Kuznetsov B.A., in: Surface phenomena in melts, Kiev, Naukova dumka, 1968, 187–191.

192. Predel B. and Eman A., *J. Less-Common Metals*, 1969, **18**, 385 – 397.
193. Eremenko V.N., *et al.*, in: Surface phenomena in melts, Kiev, Naukova dumka, 1963, 165–167.
194. Badawi W.A., *et al.*, *Bull. Chem. Soc. Japan,* 1990, **63**, No.10, 2956–2960.
195. Badawi W.A., *et al.*, *Bull. Chem. Soc. Jap.*, 1990, **63**, No.6, 1795–1800.
196. Vol A.E. and Kagan I.K., The structure and properties of binary metallic systems, vol.3, Moscow, Nauka, 1976, 494–507.
197. Okajima B.K. and Sakao J., *Trans. Jap. Institute of Metals*, 1968, **9**, No.4, 325–330.
198. Terpilovski I. and Przezdziecka E., *Archiw. Hutnictwa,* 1960, No.3. 281–289.
199. Hirth I.P., Thermodynamics of surfaces, Structure and Properties, Metal surfaces, Tokyo, 1973, 10–33.
200. Eriksson I., *Adv. Chem. Phys.*, 1964, **6**, 145–175.
201. Yatsenko S.P., Gallium. Interaction with metals, Moscow, Nauka, 1974.
202. Ibragimov Kh.I., in: Physics of surface phenomena in melts, Groznyi, Chech. Inzh. Knizhn. Izd-vo, 1977, 59–65.
203. Naidich Yu.Z. and Perevertailo V.M., in: Physical chemistry of the surface of melts, Tbilisi, Metsniereba, 1977, 23–30.
204. Ashkhotov O.G., et al., Dokl. AN SSSR, 1984, **274,** No.6, 1349–1352.
205. Morachevskii A.G. and Sladkov I.B., Thermodynamic calculations in metallurgy, Moscow, Metallurgiya, 1985.
206. Morachevskii A.G. (ed.), Thermodynamics of the liquid–vapour equilibrium, Moscow, Khimya, 1989.
207. Koval'chuk V.F. and Palova G.D., Investigation of the free surface energy of liquid metallic alloys of the indium–thallium system, Dep. ONIITE, Khim. Cherkassy, No. 496/75, 1975.
208. Kuznetsov B.V., in: Surface phenomena in metals and alloys and the role of the phenomena in the processes of powder metallurgy, Kiev, Publishing House of the Academy of Sciences of the Ukrainian USSR, 1961, 211–221.
209. Nizhenko V.I., in: Adhesion and contact interaction of melts, Kiev, Naukova Dumka, 1988, 3–17.
210. Eremenko V.N. and Vasiliu M.N., *Ukr. Khim. Zh.,* 1972, **38**, No.2, 118–121.
211. Dadashev R.Kh., Ibragimov Kh.I., and Yushaev S.M., *Izv. VUZ, Tsvetnaya Metallurgiya*, No.1, 121–122.
212. Hultgren R., *et al.*, Selected values of thermodynamic properties of metals and alloys, London, 1963, 926–929; 937–942.
213. Khokonov Kh.B., et al., Zh. Izv. SKNTs VSh. Estestv. Nauki, 1973, No.2, 60–65.
214. Khokonov Kh.B., Investigation of the surface tension and work function of the electron of the metal and alloys in the liquid and solid state, PhD thesis, Nal'chik,
215. Ibragimov Kh.I., in: Physics of interfacial phenomena, Nal'chik, KGBU, 1985, 44–57.
216. Khantadze D.V., Topuridze N.I., in: The physics of surface phenomena in melts, Groznyi, ChIGU, 1978, 11–16.
217. Khokonov Kh.B., *et al.*, *DAN SSSR*, 1973, **210**, No.4, 899–902.
218. Ibragimov Kh.I. and Nal'giev A.R.-M., in: Physical chemistry of the interface of contacting phases, Kiev, Naukova Dumka, 1976, 85–90.
219. Ibragimov Kh.I., *et al.*, in: Surface phenomena in melts and solid phases formed from them, Nal'chik, Kab.-Balk. Kn. Izd-vo, 1965, 269–277.
220. Sagov B.B. and Ibragimov Kh.I., in: The interface boundary and its properties, Naukova dumka, Kiev, 1980, 56–60.
221. Yatsenko S.P., Indium. Properties and application, Moscow, Nauka, 1987.

222. Lashko N.F. and Lashko S.V., Braizing of metals, Moscow, Mashinostroenie, 1977.
223. Zedgenidze I.G., Experiment design in the investigation of multicomponent systems, Moscow, Nauka, 1976.
224. Lazarev V.B., in: The surface phenomena in melts, Kiev, Naukova Dumka, 1968, 170–173.
225. Konyukhova N.P., *et al.*, : The physical chemistry of surface phenomena in melts, Kiev, Naukova Dumka, 1971, 64–67.
226. Doge G., Z., *Naturforschung*, 1966, **215**, No.3, 276–284.
227. Masdesi T., *et al.*, Bull. of the Institute of Atomic Energy, Kyoto Univ., 1975, **47**, No.3. 37–39.
228. Pokrovskii N.L., *et al.*, *DAN SSSR*, 1968, **181**, No.1, 80–83..
229. Masdesi T., *et al.*, Bulletin of the Institute of Atomic Energy Kyoto Univ., 1975, **47**, No.3, 37–39.
230. Abdel-Aziz Abol-Hassan and K.Kirahon Hohonmad, *Zeitschrift fur Metallkunde*, 1977, **68**, No.6, 437–439.
231. Pokrovskii N.L., *et al.*, *Zh. Fiz. Khim.*, 1969, **48**, No.8, 2158–2159.
232. Berthon R.E., Toygas R., *Met. Trans.*, 1970, No.10, 2978–2979.
233. Yatsenko S.P., *et al.*, in: Properties of gallium in aqueous solutions and melts, The Proceedings of the Institute of Chemistry of the UF AN SSSR, Sverdlovsk, 1966, No.12, 66–89.
234. Vecher A.A., et al., *Vestn. Belorus. Gosuniversiteta, Ser. 2, Khimiya*, 1974, No.3, 3–6.
235. Klym N.M., *et al.*, in: The structure and properies of metallic and slag melts, Seventh national conference, Chelyabinsk, 1990, v.2, part 1, 108–109.
236. Adler Yu.G. and Grabovskii Yu.B., *Zavod. Lab.*, 1977, No.10, 1253–1259.
237. Chemleva T.A. and Mikeshina N.G., in: New concepts in the experiment design, Moscow, Nauka, 1969.
238. Kararov V.V., Cybernetic methods in chemistry and chemical technology, Moscow, Khimiya, 1971.
239. Goitemirov R.U., Investigations of the surface properties of multicomponent systems using mathematical experiment design, Thesis, Groznyi, ChIGU, 1987.
240. Zedgenidze I.G., *et al.*, in: Using mathematical methods for investigations of multicomponent systems, Moscow, Metallurgiya, 1974, 3–11.
241. Scheffe H., *J. Royal Statistical Society*, 1963, **25**, No.2, 235.
242. Goitemirov R.U., *et al.*, in: Cybernetics methods in chemistry and chemical technology, Proceedings of the Second national seminar of young scientists, Groznyi, 1984, 73–74.
243. Dadashev R.Kh., Ibragimov Kh.I., in: Physics of interfacial phenomena, Nal'chik, KBGU, 1978, 32–40.
244. Shakhparonov M.I., Introduction into the advanced theory of solutions, Moscow, Vysshaya shkola, 1976.
245. Pavlov V.V. and Popel' S.I., *Zh. Fiz. Kim.*, 1970, **44**, No.5, 1190–1194
246. Ibragimov Kh.I. and Sagov B.B., in: Physics of interfacial phenomena, Nal'chik, KGBU, 1976, 41–47.
247. Ofitserov A.A., *et al.*, *Izv. VUZ, Chernaya Metallurgiya*, 1966, No.9, 5–10.
248. Kononenko V.I., *et al.*, *Zh. Fiz. Kim.*, 1972, **46**, No.6, 1589–1590.
249. Alchagirov B.B., *et al.*, in: Physics of interfacial phenomena, Nal'chik, KGBU, 1976, 42–52.
250. Sukhman A.L., in: Physical–chemical investigations of metals and alloys, Proceedings of the Institute of Chemistry of the Uralsk Scientific Centre of the USSR

Academy of Sciences, Sverdlovsk, 1974, No.29, 53.

251. Lazarev B.P. and Semenchenko V.K., *Izv AN SSSR, Otd. Khim. Nauk*, 1962, No.11, 2082–2085.

252. Kohler P., *Z. Chem.*, 1965, **96**, 1228–1251.

253. Pugachevich P.P. and Lazarev V.B., *DAN SSSR*, 1957, No.3, 445–447.

254. Lazarev V.B., *Zh. Neorg. Khimii*, 1958, No.10, 2318–2338.

255. Dadashev R.Kh..Thermodynamics of surface phenomena,a textbook, Groznyi, ChIGU, 1988.

256. Dzhemilev I.K., *et al.*, in: *Zh. Fiz. Khim.*, 1966, **41**, No.1, 47–51.

257. Dzhemilev I.K., *et al.*, in: Surface phenomena in melts and solid phases formed from them, Nal'chik, Kab-Bal. Kh. Izd., 1965, 306–311.

258. Otsiferov A.A., Surface tension of ternary metallic solutions, Thesis, Moscow, 1967.

259. Backes H.M., *et al., Chem. Eng. Sci.*, 1990, **45**, No.1, 275–286.

260. Savvin V.S. and Ibragimov Kh.I., in: Physics of surface phenomena, Groznyi, 1977, 172–184.

261. Kozin L.F., Amalgam pyrometallurgy, Nauka, Alma–Ata, 1973.

262. Popel' S.I. and Kozhurkov V.N., *Metally*, 1974, No.2, 49–52.

263. Zhukov A.A., Surface properties of separated melts based on iron and silver, Thesis, Sverdlovsk, 1975.

264. Popel' S.I., *et al.*, in: Physical and chemical investigations of metallurgical processes, Sverdlovsk, UPI, 1976, 43–48.

265. Mit'ko M.M., *et al., Izv. VUZ, Tsvetnaya Metallurgiya*, 1979, No.1, 91–94.

266. Mit'ko M.M., *et al.*, in: Surface properties of melts, Kiev, Naukova dumka, 1982, 122–127.

267. Zakharova T.V., *et al., Izv. VUZ, Tsvetnaya Metallurgiya*, 1979, No.1, 86–90.

268. Cao W. and Han S., *Izhejiang Univ.*, 1990, **24**, No.1, 93–99.

269. Sukhman A.L., *et al., Izv. VUZ, Metally,* 1972, No.6, 94–95.

270. Kononeko V.I., *et al.*, in: The physical–chemical investigations of liquid metals and alloys, Sverdlovsk, 1974, 62–68.

271. Ofitserov A.A., *DAN SSSR*, 1971, No.4, 91–92

272. Ibragimov Kh.I., *et al.*, in: Heat conductivity and diffusion in technological processes, Riga, 1977, 29–32.

273. Naidich Yu.V., *et al.*, Zh. Fiz. Khim., 1984, **58**, Nio.5, 1810–1811.

274. Naidich Yu.V., *et al.*, in:Adhesion and brazing of materials, 1988, No.20, 1–4.

275. Goitemirov R.U. and Ibragimov Kh.I., in: Proceedings of the Conference for the Scientific results of works in 1982, ChIGU, Groznyi, 1983, 230–231.

276. Dadashev R.Kh., in: Proceedings of the Conference for the Scientific results of works in 1982, ChIGU, Groznyi, 1983, 225–227.

277. Iond I.C, *et al., J. de Chimie Physique et de Physico-chimie Biologique*, 1974, **1**, No.1–8, 1113–1220.

278. Naidich Yu.V., *et al.*, Metally, 1988, No3, 57–60.

279. Dadashev R.Kh., *et al.*, in: Current problems in the mechanics of liquids and gas, Groznyi, 1986, 113.

280. Dadashev R.Kh., *et al.*, in: Methods of cybernetics in chemistry and chemical technology, Groznyi, 1984, 26–28.

281. Sukhman A.L., *et al., Metally*, 1988, No.2, 52–53.

282. Morachevskii A.G., in: Thermodynamic properties of metallic alloys and advanced methods of investigating them, Kiev, Naukova Dumka, 1976.

283. Morachevskii A.G., *Zh. Fiz. Khim.*, 1968, **42**, No.3, 573–584.

284. Worch E. and Kummel R. Zur prognose des Gemischadsorptionsverhaltens, Wiss. z. Techn. Hochhsch. Carl, Schorlerner Leuna, L'erseburg 1988, Bd.30, No.1, S.36–

41.

285. Anikeev E.F., *et al.*, in: Adhesion of melts and brazing of materials, 1986, No. 17, 79–82.

286. Ustinov E.F., *Zh. Fiz. Khim.,* 1981, **62**, No.6, 1441–1454.

287. Howell W.I. and Eckert C.A., *Z.Metallk.*, 1990, **8**,No.5, 335–340.

288. Dadashev R.Kh., Ibragimov Kh.I. and Savvin V.S., in: Surface properties of melts, Kiev, Naukova dumka, 1982, 7–11.

289. Mit'ko M.M., *et al.*, in: Surface properties of melts, Kiev, Naukova dumka, 1982, 122–127.

290. Chegodaev A.I., *et al.*, in: Proceedings of the Second national conference on Surface phenomena in liquids, Leningrad, Leningrad State University, 1978, 195–196.

291. Mit'ko M.M. and Chegodaev A.I., in: Proceedings of the Second scientific and technical conference of young scientists of the Uralsk region, Sverdlovsk, UPI, 1977, 82–83.

292. Ukhov V.F., *et al.*, in: Physical chemistry of metallurgical melts, Sverdlovsk, 1962, 4.2, 80–20.

293. Petrushevskii M.S., *Izv. VUZ, Chernaya Metallurgiya,* 1978, No.8, 18–20.

294. Shchergin L.M., *et al.*, in: Physical chemistry of surface phenomena in melts, Kiev, Naukova dumka, 1971, 161–166.

295. Zadumkin S.N. and Semenchenko V.K., *Zh. Neorgan. Khim.*, 1960, **5**, No.8, 892–893.

296. Korol'kov A.M., *Izv. AN SSSR, Otd. Tekhn. Nauk*, 1956, No.2, 35–42.

297. Nizhenko V.N. and Eremenko V.N., Poroshk. Metall., 1964, No.2, 11–18.

298. Popel' S.I., Surface phenomena in melts, Moscow, Metallurgiya, 1994.

299. Ibragimov Kh.I., Thermophysical properties of metallic and polymer melts, Moscow, Intermet Engineering, 2006, 287.

300. Mekhdiev I.G., *Zh. Fiz. Khim.*, 2001, **75**, No.4, 749–751.

301. Mekhdiev I.G., *Metallofizika i Noveishie Tekhnologii,* 2000, **22**, No.5, 3–6.

302. Storonkin A.V., Thermodynamics of heterogeneous systems, Leningrad, Leningrad State University, 1967.

303. Dadashev R.Kh., *Rasplavy*, 1992, No.6.

304. Dadashev R.Kh., Surface tension and adsorption in multicomponent melts, Thesis, Ekaterinburg, 1993.

305. Rusanov A.I., Tension measurements, Leningrad, Khimiya, 1974.

306. Alchagirov B.B., and Dadashev R.Kh., The sessile droplet method used for the determination of the density and surface tension of metals and alloys, a textbook, Nal'chik, KBGU, 1975.

307. Karamurzov B.S., Surface tension, density and work function of the electron of low-melting binary systems based on gallium, Thesis, Nal'chik, KBGU, 1975.

308. Dadashev R.Kh.. in: The proceedings of the all-Russian scientific conference, Groznyi, 2003, 82–86.

309. Dadashev R.Kh., in: The proceedings of the all–Russian scientific conference: Physics of interfacial phenomena and processes of interaction of energy fluxes with solids, Nal'chik, KBGU, 1995.

310. Arsemikov U.V., *et al.*, in: Adhesion of melts and brazing of materials, Kiev, Naukova dumka, 1971, 23.

311. Dadashev R.Kh.., Kutuev R.A., Ibragimov Kh.I., in: Russian conference: Heat and mass transfer and properties of liquid metals, Obninsk, 2002, 105–107.

312. Goicochea J., et al., *J. Mater. Sci.*, 1992, **27**, No.19, 5247–5252.

313. Lee Hond-Kee, *et al., ISIJ International*, 1993, **33**, No.8, 833–838.

314. Taova T.M., in: Proceedings of the Ninth international symposium: Ordering in metals and alloys, Rostov-on-Don, 2006, 241–245.

315. Arkhestov R.Kh., *et al.*, *Vestnik AN ChR*, Groznyi, 1999, No.2, 106–111.

316. Arkhestov R.Kh., *et al.*, in: Physics and chemistry of advanced materials, Nal'chik, KBGU, 1998, 10–14.

317. Moser Z., *et al.*, *J. Electron. Mater.*, 2001, **30**, No.9, 1109–1111.

318. Kucharski M. and Gluzinska M., Surface tension of multicomponent solutions: Arch. Technol. Masz. Autom., 2001, 21, Num. spec. 123–130. Kucharski M., *et al.*, Surface tension of Cu–Ag–Au alloys, *J. Univ.Chem. Technol. and Met.*, 2002. **37**, No.3, 93–100.

319. Sagov B.B. and Ibragimov Kh.I., in: Phase boundaries and their properties, Kiev, Naukova dumka, 1980, 56–60.

320. Sagov B.B., *et al.*, *Rasplavy,* 1993, No.2.

321. Dadashev R.Kh., *et al.*, Surface properties of indium–tin–lead, indium–tin–gallium, and thallium–lead–bismuth melts, 24.07.84, dep. VINITI, No. 5356-84.

322. Dadashev R.Kh., *et al.*, Surface properties of indium–tin–thallium melts, dep. VINITI, 27.09.85, No.6927–85.

323. Dadashev R.Kh., *et al.*, in: Proceedings of the National seminar on the Magnetism of rare earth alloys, Groznyi, 1988.

324. Dadashev R.Kh. and Popel' S.I., *Zh. Fiz. Khim.*, 1992, No.6. in: proceedings of the National seminar on the magnetism of rare earth alloys, Grozny, 1988.

325. Dadashev R.Kh., *et al.*, *Rasplavy*, No.6, 1992.

326. Dadashev R.Kh., *et al.*, *Rasplavy*, No.1, 1993.

327. Dadashev R.Kh., *et al.*, *Rasplavy*, No.4, 1994.

328. Dadashev R.Kh., *Rasplavy*, No.6, 1994.

329. Dadashev R.Kh. and Ibragimov Kh.I., in: Adhesion of melts and brazing of materials, Kiev, Naukova dumka, 1994.

330. Dadashev R.Kh. and Kutuev R.A., in: Proceedings of the International seminar: Thermophysical properties of substances, Nal'chik, KBGU, 2001.

331. Dadashev R.Kh. and Kutuev R.A., in: Proceedings of the all-Russian scientific conference, Groznyi, December 2002, 96–105.

332. Dadashev R.Kh., *et al.*, in: Proceedings of the all-Russian scientific conference, Groznyi, December 2002, 105–115.

333. Dadashev R., *et al.*, in: Fourth international conference: High-temperature capillarity HTC–2004.

334. Dadashev R., *et al.*, in: Fourth international conference: High-temperature capillarity HTC–2004.

335. Dadashev R.Kh., *et al.*, *Zh. Fiz. Khim.*, 2007, **81**, No.6, 1–3.

336. Dadashev R.Kh., *et al.*, in: Proceedings of the Third Russian scientific and technical conference: Physical properties of metals and alloys, Ekaterinburg, 2005, 96.

337. Dadashev R.Kh., *et al.*, in: Thermophysical properties of substances and materials, Proceedings of the Ninth Russian conference, vol.1, St. Petersburg, 2005, 64.

338. Dadashev R.Kh.., *et al.*, in: Thermophysical properties of substances and materials, Proceedings of the Ninth Russian conference, vol.1, St. Petersburg, 2005, 65.

339. Dadashev R.Kh. and Elimkkhanov D.Z., *et al.*, in: Thermophysical properties of substances and materials, Proceedings of the Ninth Russian conference, vol.1, St. Petersburg, 2005, 64–68.

340. Dadashev R.Kh., *et al.*, in: Proceedings of 9th Interdisciplinary international symposium: Phase transformations in solid solutions and alloys, OMA-9, Rostov-on-Don, 2006.

Index